INVESTIGATIONS IN NONLINEAR OPTICS AND HYPERACOUSTICS

ISSLEDOVANIYA PO NELINEINOI OPTIKE I GIPERAKUSTIKE

ИССЛЕДОВАНИЯ ПО НЕЛИНЕЙНОЙ ОПТИКЕ И ГИПЕРАКУСТИКЕ

The Lebedev Physics Institute Series

Editor: Academician D. V. Skobel'tsyn

Director, P. N. Lebedev Physics Institute, Academy of Sciences of the USSR

Proceedings (Trudy) of the P. N. Lebedev Physics Institute

Volume 58

INVESTIGATIONS IN NONLINEAR OPTICS AND HYPERACOUSTICS

Edited by
Academician D. V. Skobel'tsyn
Director, P. N. Lebedev Physics Institute
Academy of Sciences of the USSR, Moscow

Translated from Russian by Albin Tybulewicz
Editor, Soviet Physics - Semiconductors

CONSULTANTS BUREAU
NEW YORK – LONDON
1973

Library of Congress Cataloging in Publication Data

Main entry under title:

Investigations in nonlinear optics and hyperacoustics.

(Proceedings (Trudy) of the P. N. Lebedev Physics Institute, v. 58)
Translation of Issledovaniya po nelineinoi optike i giperakustike.
Includes bibliographical references.
1. Nonlinear optics. 2. Light–Scattering. 3. Ultrasonics. I. Skobel'tsyn, Dmitriĭ
Vladimirovich, 1892-   ed. II. Series: Akademiiā nauk SSSR. Fizicheskiĭ institut.
Proceedings, v. 58.
QC1.A4114 vol. 58 [QC446.2]   530'.08s [535]   73-79425
ISBN 978-1-4684-7804-4   ISBN 978-1-4684-7802-0 (eBook)
DOI 10.1007/978-1-4684-7802-0

The original Russian text was published by Nauka Press in Moscow in 1972 for the
Academy of Sciences of the USSR as Volume 58 of the Proceedings of the P. N.
Lebedev Institute. The present translation is published under an agreement with
Mezhdunarodnaya Kniga, the Soviet book export agency.

CONTENTS

SPECTRA OF THE THERMAL AND STIMULATED SCATTERING OF LIGHT IN THE WING OF THE RAYLEIGH LINE *

G. I. Zaitsev

The spectra of the thermal scattering of light by anisotropy fluctuations (corresponding to the wing of the Rayleigh line) and the profiles of the depolarized lines of the Raman scattering in a liquid were investigated and compared. An asymmetry was observed in the indicatrix of the thermal scattering in the wing of the Rayleigh line of some liquids (for example, benzene or toluene). In the region beginning at ~ 20 cm^{-1} from the frequency of the exciting line the scattering intensity was found to be higher in the forward than in the backward direction. This was attributed to the existence of short-lived complexes, which had the short-range orientational order. A ruby laser of ~ 100 MW output was used as the excitation source in a study of the temperature dependence of the Stokes stimulated scattering in the wing of the Rayleigh line. A nondegenerate nonlinear four-photon interaction gave rise to the Stokes and the anti-Stokes wings in the stimulated spectrum. Such stimulated scattering was observed at a small angle ($\sim 2°$) with respect to the direction of propagation of the exciting light, whereas at large scattering angles only the Stokes wing was observed. These results were in agreement with the theory of stimulated scattering.

INTRODUCTION

Molecular scattering of light is a wide-ranging part of molecular optics and it is closely related to various aspects of molecular physics and molecular acoustics. Some of the important problems in physics have been solved by the molecular scattering method [1, 2].

The phenomenon of molecular scattering of light transmitted by a medium results from fluctuations in the refractive index of this medium. Such fluctuations are due to fluctuations of the density of the medium and of the orientation of anisotropic molecules (in the case of pure substances) or fluctuations of the concentration (in the case of homogeneous solutions). The principal characteristics of the scattered light, i.e., its intensity, polarization, and spectral composition, reflect the properties of the scattering medium [1].

Spectroscopic investigations of the scattered light are of great scientific importance because they provide one of the most effective methods for the study of the kinetics of processes occurring in various media. The spectral composition of the scattered light is much richer than that of the exciting light because of the modulation resulting from the variation of the fluctuations with time. Every kind of thermal fluctuation is reflected in the scattered-light

---

*Condensed text of a thesis submitted for the degree of Candidate of Physicomathematical Sciences, defended on February 26, 1968, at the P. N. Lebedev Physics Institute, Academy of Sciences of the USSR. Scientific supervisors: I. L. Fabelinskii and V. S. Starunov.

spectrum: the slowest isobaric fluctuations of the density and the concentration give rise to the central line, whereas adiabatic fluctuations of the density are responsible for the displaced (Brillouin*) components of the fine structure of the Rayleigh line [1-4]. The spectrum of the light scattered by liquids includes, in addition to the fine structure components, a continuous depolarized background with an intensity maximum which coincides with the center of the undisplaced line. This background extends in both directions up to 200 cm^{-1} from the displaced line and it is known as the wing of the Rayleigh line [5, 6]. This wing is the result of modulation of the scattered light by fast changes in the fluctuations of the anisotropy [3, 4].

Investigations of the scattered-light spectra have provided very extensive information on the thermal motion of molecules in various media and on some properties of these media [1, 7-10] in the high-frequency range (10^{10}-10^{13} Hz) where other physical methods are unsuccessful. Studies of the fine structure spectrum have yielded values of the velocity and the absorption coefficient of hypersound in liquids and, consequently, the dispersion of the velocity of sound and the relaxation time of the bulk viscosity. Some valuable information on the thermal orientational motion of molecules in low-viscosity liquids can be deduced from the distribution of the intensity in the wing of the Rayleigh line.

An analysis of the experimental data on the distribution of the intensity in the wing of the Rayleigh line, based on the modern theory of this wing [8, 11, 12], yields the lifetimes of molecules in potential wells in a liquid and the frequency of elastic rocking of molecules in the field of its neighbors. Moreover, different parts of the wing of the Rayleigh line give the anisotropy relaxation time, the internal friction, and the elastic moduli at $\sim 10^{12}$-10^{13} Hz. Consequently, one can utilize the scattering of light in studies of the relaxation of the internal friction in liquids. The information on the behavior of liquids at these very high frequencies which can be deduced from the molecular scattering phenomena is quite unique.

The use of ruby lasers, capable of emitting light of unheard-of intensity, in studies of the interaction of light with matter has made it possible to discover new optical phenomena and to obtain fresh information on the known effects.

In the thermal scattering of light a weak field of the exciting light wave has a negligible effect on the medium. The situation changes drastically when a giant ruby-laser pulse is used. In this case the electric field of the incident wave is so high that, in combination with the scattered-wave field, it produces considerable striction and orientation-inducing forces. The strictional force gives rise to the stimulated Brillouin scattering which was first observed by American physicists [13-15]. The stimulated scattering of light in the wing of the Rayleigh line resulting from the action of the orientation force on anisotropic molecules in a liquid was discovered by Mash, Morozov, Starunov, and Fabelinskii [16]. More recently, Kyzylasov, Starunov, Fabelinskii, and the present author [17] found another nonlinear optical effect in the form of the stimulated temperature scattering of light in liquids. In this effect the strong exciting light emitted by a ruby laser and the weak (initially thermal) light scattered by fluctuations of the entropy interact with the medium to produce strong temperature waves. The reaction between these waves and the exciting and scattered light gives rise to "pumping" of the energy from the laser beam to the scattered light and to the temperature wave.

The stimulated Brillouin scattering is manifested by a nonlinear increase of the intensity of the fine structure components of the scattering line with increasing linear dimensions of the scattering region, provided the intensity of the exciting light exceeds a certain threshold value. Under some conditions of propagation of the scattered light in a given medium and in the presence of scattering back to the ruby laser the number of the Brillouin components may reach a few tens (this is known as the regeneratively amplified stimulated Brillouin scattering) [18].

---

* The Brillouin scattering is usually referred in the Soviet literature as the Mandel'shtam-Brillouin scattering.

At large scattering angles the spectrum of the stimulated wing of the Rayleigh line is observed only on the Stokes side of the exciting line and the intensity maximum is shifted from this line by a frequency interval equal to the reciprocal of the anisotropy relaxation time of the liquid in question. However, at low scattering angles a four-photon interaction may give rise also to the anti-Stokes part of the spectrum.

There are still many unresolved problems in molecular (thermal and stimulated) scattering of light in matter existing in various states. The present paper describes the results of studies and analyses of some of these problems on the basis of the results deduced from the spectra of the thermal and stimulated scattering of light corresponding to the region of the wing of the Rayleigh line of liquids. Studies of the thermal scattering of light included investigations of the wing of the Rayleigh line observed in high-viscosity liquids, of the distribution of intensity in this wing as different scattering angles, and of the profiles of depolarized Raman lines in a wide range of frequencies. The purpose of studies of the stimulated scattering was to determine the influence of temperature on the stimulated wing of the Rayleigh line and to determine the possibility of detecting the four-photon interaction in the stimulated scattering corresponding to the wing region.

High-speed apparatus, which included a DFS-12 spectrometer and an improved electronic circuit, made it possible to determine experimentally the distribution of intensity in the wing of the Rayleigh line of the following three viscous liquids: salol (phenyl salicylate), benzophenone, and triacetin [19]. This investigation was carried out at various temperatures and it yielded information on the behavior of the internal friction at $\sim 10^{12}$-10^{13} Hz as well as the numerical values of the limiting frequencies of the rocking oscillations of salol, benzophenone,, and triacetin molecules (Chap. III).

Investigations of the distribution of intensity in the wing of the Rayleigh line and in the spectrum of the depolarized Raman scattering lines, first carried out by Zaitsev and Starunov [20] on pure liquids in a wide range of frequencies, undoubtedly helped in the understanding of the nature of broadening of the Rayleigh and the Raman scattering lines. A comparison of the profiles of the Rayleigh and the Raman lines indicated that in the case of liquids whose molecules had three identical principal moments of inertia the nature of the intensity distribution was the same in the frequency range 20-80 cm^{-1}. In the case of molecules which had different moments of inertia the spectral distributions were different for the Raman lines and the far parts of the wing of the Rayleigh line. The reason for this difference is discussed in Chap. III.

Somewhat unexpected results were obtained in an investigation of the angular distribution of the intensity in the spectrum of the wing of the Rayleigh line [21] of benzene, toluene, carbon disulfide, carbon tetrachloride, salol, and benzophenone. It was established that in the case of carbon disulfide, carbon tetrachloride, benzophenone, and salol at room temperature the indicatrix was symmetrical for all the frequencies in the spectrum of the wing, as expected from the theory of scattering of light [1]. However, in the case of benzene, toluene, and salol at 120° the wing was dominated, beginning from some frequency, by the forward scattering. This result was attributed to the existence of very-short-lived orientational order and of correlation in the orientational motion of molecules in the liquids (Chap. III).

The results of an investigation of the spectra of the stimulated molecular scattering of light are given in Chap. IV. The stimulated wing of the Rayleigh line was studied for the first time in five new liquids. For some of these liquids the studies were carried out at various temperatures and different conditions of propagation of light in the scattering medium and in the ruby laser. The origin of some of the Brillouin components was determined [22]. Moreover, the four-photon interaction between the exciting, Stokes, and anti-Stokes waves on the one hand and the nonlinear medium on the other was observed (at low angles) for the first time [23] in the stimulated scattering of light corresponding to the wing of the Rayleigh line.

THEORETICAL AND EXPERIMENTAL INVESTIGATIONS OF THE SPECTRA OF THE THERMAL AND STIMULATED SCATTERING OF LIGHT

§1. Spectral Composition of the Light Scattered by Anisotropy Fluctuations in Liquids

The first spectra investigations of the light scattered in liquids were carried out by Cabannes and Daure [5] and by Raman and Krishnan [6]. These investigations demonstrated that a continuous and fairly strong background, which was absent in the spectrum of the exciting light, extended on both sides of the Rayleigh line. This background, which is typical of all liquids, is known as the wing of the Rayleigh line.

The nature of the intensity distribution in the spectrum of such a wing is fairly complex [10, 24]: near the undisplaced (central) line the intensity decreases quite rapidly with increasing frequency but further away the fall of the intensity slows down and this is followed by a second region of rapid fall in the intensity. Moreover, a considerable depolarization is observed in the wing of the Rayleigh line, as established in the very first investigations.

We shall not discuss in detail the various hypotheses relating to the origin of the wing of the Rayleigh line [6, 25-28], except for pointing out that the theory of Landau and Placzek [29] was found to be most fruitful. According to Landau and Placzek, the broadening of the Rayleigh scattering line is due to relaxation processes in the liquid in question and the width of the wing is governed by the reciprocal of the relaxation time of the dipole moment.

Leontovich [3] developed a quantitative phenomenological theory of the scattering of light. According to this theory the wing is due to the scattering by fluctuations of anisotropy of the liquid being considered and the spectral composition is governed by the law which describes the dispersal of these fluctuations:

$$\dot{S}_{ik} + \frac{1}{\tau} S_{ik} = \dot{e}_{ik} - \frac{1}{3} e_{ii} \delta_{ik}. \tag{1}$$

Here, S_{ik} is the anisotropy tensor describing the deviation of the axes of anisotropic molecules from the random positions; e_{ik} is the strain tensor; τ is the relaxation time of the anisotropy of the medium. Leontovich used Eq. (1) to obtain an expression for the distribution of the intensity in the wing, which is valid far from the Brillouin components:

$$I(\Omega) = \text{const} \frac{kT}{\mu} \frac{\tau}{1 + \Omega^2 \tau^2}, \tag{2}$$

where μ is the shear modulus and $\tau = \eta / \mu$ (η is the viscosity of the medium). Leontovich suggested also that if a liquid can be described by the Debye model [30], the anisotropy relaxation time is given by the expression

$$\tau = \frac{1}{3} \tau_D = \frac{4\pi r^3 \eta}{3kT}. \tag{3}$$

Here, τ_D is the relaxation time of the dipole moment in a polar liquid and r is the radius of the molecules. It follows from Eq. (2) that the half-width of the wing of the Rayleigh line is related to the anisotropy relaxation time:

$$\Delta\nu = \frac{1}{\pi c\tau} \, , \, \mathrm{cm}^{-1} \tag{4}$$

The experimental results [10, 12, 24, 31] indicate that the spectral distribution of the intensity in the wing cannot be described by the simple dispersion formula (2) because the second derivative of the anisotropy tensor is missing from the relaxation equation (1) and this is permissible only up to frequencies of $\sim 10^{12}$ Hz. It follows that Eq. (2) is unsuitable for the description of the high-frequency part of the wing of the Rayleigh line. Therefore, an allowance must be made for inertia in the intensity distribution.

This allowance was made by Zhivlyuk [32], who replaced Eq. (1) with

$$a\ddot{S}_{ik} + \dot{S}_{ik} + \frac{1}{\tau} S_{ik} = \dot{e}_{ik} - \frac{1}{3} e_{ii}\delta_{ik}, \tag{5}$$

where a is a parameter representing the inertia.

Zhivlyuk used Leontovich's approach in the calculation of the distribution of the intensity in the wing of the Rayleigh line and he obtained the following expression for the intensity in the wing:

$$I(\Omega) = \mathrm{const} \, \frac{kT}{\mu} \, \frac{\tau}{(1 - a\Omega^2)^2 + \Omega^2\tau^2} \, , \tag{6}$$

which reduces to Leontovich's formula (2) if $a = 0$.

An allowance for the inertia in the relaxation equation (1) makes it possible to explain qualitatively the rapid fall of the intensity in the far part of the wing. However, it should be noted that the viscosity in Eq. (6) — introduced via τ — is presumed to be frequency-independent. Investigations of the temperature dependence of the intensity in the spectrum of the wing of the Rayleigh line, carried out by Fabelinskii [10, 31], demonstrated that the rate of narrowing of the far region is much less than the rate of rise of the viscosity of the liquid. An analysis of this experimental observation led Fabelinskii to the idea that at frequencies corresponding to the far part of the wing ($\sim 10^{13}$ Hz) the shear viscosity may undergo relaxation.

Rytov [4] developed a general correlation theory and made allowance for the relaxation of the viscosity without recourse to any specific assumptions about the nature of dispersion. According to Rytov, a liquid (of any viscosity) or an amorphous solid has two complex elastic moduli, scalar thermal parameters, and scalar elasto-optical and thermodynamic properties. Rytov [4] considered only those internal processes in the scattering medium which are associated with strain and temperature fluctuations.

Rytov's theory is not thermodynamic but it is based on correlation statistics. Consequently, it can be applied when any of the parameters of the theory are frequency-dependent.

Rytov assumed that fluctuations are small and restricted his treatment to the case of a linear relationship between the deviation of the permittivity from its average value and the strain tensor e_{ik}:

$$\Delta\varepsilon_{ik} = X(\Omega)\left(\dot{e}_{ik} - \frac{1}{3} e_{ii}\delta_{ik}\right).$$

Next, making allowance for the relaxation of the viscosity and the shear modulus, he expressed the complex shear modulus in the form

$$\bar{\mu} = \mu(\Omega^2) + i\Omega\eta(\Omega^2) = \frac{\mu_\infty i\Omega}{i\Omega + \frac{1}{\tau_\eta}} \, . \tag{7}$$

It follows that $\eta(\Omega^2) = \eta_0/(1 + \Omega^2\tau_\eta^2)$ and $\tau_\eta = \eta_0/\mu_\infty$. Rytov finally obtained the following expression for the distribution of the intensity in the wing of the Rayleigh line:

$$I(\Omega) = \text{const } kT \left| \frac{X(\Omega)}{\overline{\mu}(\Omega)} \right| \eta(\Omega^2). \tag{8}$$

In this theory the function X(Ω) is undefined but it can be determined if a specific model is assumed. It should be stressed that Eq. (8) applies only to that part of the scattering process which is associated solely with fluctuations of the strain tensor. Ginzburg [33] pointed out that in the case of low-viscosity liquids the scattering or fluctuations of the strain tensor is only one of the components of the general scattering or anisotropy fluctuations.

A more complete description of the spectral distribution of the intensity in the wing of the Rayleigh line is given by the semiphenomenological theory developed by Starunov [8, 11, 12]. Starunov used Leontovich's hypothesis on the relaxation origin of the wing [3] and Fabelinskii's idea on the influence of the relaxation of the viscosity on the spectral composition of the depolarized light. He also used Gross's suggestion of a connection between the wing of the Rayleigh line and the rotational rocking of molecules in a liquid [34]. However, he made no assumption about the existence of quasicrystalline regions or of short-range orientational order.

In the simplified picture of molecular motion it is assumed that the Brownian rotation of a molecule in a liquid can be formally divided into two processes: 1) orientational jumps from one equilibrium orientation to another (rotational diffusion); 2) aperiodic rotational rocking in the intervals between such jumps.

When light travels across a liquid consisting of anisotropic molecules, both types of such rotational motion will modulate the scattered light. The slower orientational jumps from one potential well to another dominate the region of the wing close to the undisplaced line whereas the second (faster) type of motion is more important in the far region of the wing.

The distribution of intensity in the spectrum of the near region of the wing, governed by the rotational diffusion, was calculated by Starunov for a linear molecule which has only one moment of inertia I in the angle and velocity space. The orientation of such a molecule can be described by the angular coordinates ϑ and φ, which are both functions of time.

If the incident light wave is polarized along the z axis, it induces a molecular moment whose components are functions of the angles [9]:

$$S_z = \cos^2\vartheta - \frac{1}{3}, \qquad S_x = \sin\vartheta \cdot \cos\vartheta \cdot \cos\varphi. \tag{9}$$

Under steady-state conditions, i.e., when the correlation functions $f_z(t) = \overline{S_z(t')S_z(t'+t)}$ and $f_x(t) = \overline{S_x(t')S_x(t'+t)}$ depend only on the time shift t, the intensity of the scattered light can be written in the form [3]:

$$\left. \begin{array}{l} I_z(\Omega) = f_z(\Omega) = \dfrac{1}{\pi}\displaystyle\int_0^\infty f_z(t)e^{-i\Omega t}\,dt, \\[4mm] I_x(\Omega) = f_x(\Omega) = \dfrac{1}{\pi}\displaystyle\int_0^\infty f_x(t)e^{-i\Omega t}\,dt. \end{array} \right\} \tag{10}$$

Starunov calculated the spectral densities of the correlation functions using the Fokker-Planck equation [35-37]. These calculations led finally to formulas describing the intensity distributions in two limiting cases. In the first case, representing a low-pressure gas ($\zeta/I \to 0$, where ζ is the internal friction), the distribution of the intensity is described by the equation*

$$I_z(\Omega) = \frac{4}{3} I_x(\Omega) = \frac{I}{120kT} \exp\left(-\frac{I\Omega^2}{8kT}\right) + \frac{1}{45} \delta(\Omega). \tag{11}$$

In the second case, when the conditions $(\zeta/I)^2 \gg 4kT/I$ and $2\zeta/I \gg \Omega$, applicable to ordinary liquids are satisfied, the corresponding expression is

$$I_z(\Omega) = \frac{4}{3} I_x(\Omega) = \frac{4}{45\pi} \frac{\zeta_0/6kT}{\left(1 - \frac{I}{4kT}\Omega^2\right)^2 + \Omega^2 \left(\frac{\zeta_0}{6kT}\right)^2}. \tag{12}$$

For most liquids the inertial term $I/4kT\Omega^2$ in Eq. (12) can be ignored, because it is smaller than unity, and therefore the spectral distribution of the intensity in the wing can be described by a dispersion formula which is identical with Eq. (2) and in which the anisotropy relaxation time is

$$\tau = \frac{\zeta_0}{6kT}. \tag{13}$$

If the internal friction is given by the Stokes formula

$$\zeta_0 = 8\pi r^3 \eta, \tag{14}$$

we find that Eq. (13) reduces to the standard expression for the anisotropy relaxation time (3). However, it is − in principle − incorrect to describe molecules by means of the hydrodynamic Stokes theory developed for the motion of a macroscopic sphere in viscous liquid. Only experience can justify this step.

Molecules in a liquid can undergo rotational diffusion as well as elastic rocking about their equilibrium orientations. If the deviation $\Delta\vartheta$ of the axis of a molecule from its equilibrium position is small, we obtain the following stochastic equation derived by Starunov:

$$I\Delta\ddot{\vartheta} + \zeta\Delta\dot{\vartheta} + g\Delta\vartheta = M(t). \tag{15}$$

Here, g is the modulus of elasticity and M(t) is a random force. Such rocking of the molecules alters the moment by an amount

$$\Delta S_z = -2\sin\vartheta\cos\vartheta\,\Delta\vartheta. \tag{16}$$

The spectral densities corresponding to the correlation functions $\overline{\Delta S_z(t')\Delta S_z(t'+t)}$ and $\overline{\Delta\vartheta(t')\Delta\vartheta(t'+t)}$, are related by the expression

$$\overline{(\Delta S_z)_\Omega^2} = \overline{4\sin^2\vartheta\cos^2\vartheta\,(\Delta\vartheta)_\Omega^2}. \tag{17}$$

The stochastic equation (15) was used by Starunov [8, 11, 12] to find $\overline{(\Delta\vartheta)_\Omega^2}$ and, after averaging in Eq. (17) over the angles, an expression for the intensity of the scattered light:

$$I_z(\Omega) = \frac{4}{3} I_x(\Omega) = \frac{4}{45\pi} \frac{6kT}{g} \frac{\zeta/g}{\left(1 - \frac{I}{g}\Omega^2\right)^2 + \Omega^2(\zeta/g)^2}. \tag{18}$$

---

* This and later expressions for the intensity do not include an unimportant factor, which is independent of Ω and ζ.

Here, ζ and g are frequency-independent. The above expression describes the same spectral region as Eq. (6).

Starunov used Fabelinskii's idea on the relaxation of the viscosity (internal friction) and made allowance for the dispersion of ζ and g, rewriting Eq. (15) in the form

$$I\Delta\ddot{\vartheta} + \zeta_0\Delta\dot{\vartheta} + g_0\Delta\vartheta = M(t), \quad t > \tau_\zeta, \left.\vphantom{\begin{matrix}a\\b\end{matrix}}\right\}$$
$$I\Delta\ddot{\vartheta} + g_\infty\Delta\vartheta = M(t), \quad t < \tau_\zeta. \qquad\qquad (19)$$

Here τ_ζ is the relaxation time of the internal friction (viscosity) and of the modulus of elasticity. The two expressions in Eq. (19) yield the formula (18) for I_z but now the parameters ζ and g depend on the frequency:

$$\zeta = \frac{\zeta_0}{1 + \Omega^2\tau_\zeta^2}, \qquad g = g_0 + \tau_\zeta\zeta_0\frac{\Omega^2}{1 + \Omega^2\tau_\zeta^2}. \qquad (20)$$

In Eqs. (19) and (20) the symbols ζ_0 and g_0 represent the low-frequency values (below the relaxation region) and g_∞ is the high-frequency value (beyond the relaxation region), given by

$$g_\infty = g_0 + \frac{\zeta_0}{\tau_\zeta}. \qquad (21)$$

If the internal friction (viscosity) relaxation time is long, it follows from Eqs. (18), (20), and (21) that a pair of depolarized lines of frequency $\Omega_{max} \approx \sqrt{g_\infty/I} = \Omega_c$ should appear on both sides of the undisplaced line. Naturally, this theory predicts only one pair of lines because only one moment of inertia of the molecule is taken into account.

Thus, Starunov's theory describes the intensity distribution throughout the wing, with the exception of a small region near the frequency $\Omega = 1/\tau_p$, where τ_p is the time spent by a molecule in a potential well. At frequencies $\Omega > 1/\tau_p$ the spectral distribution is given by Eq. (18) with the parameters set out in Eq. (20), whereas at frequencies $\Omega < 1/\tau_p$ this distribution is given by Eq. (12). It is evident from Eqs. (12), (18), and (20) that the spectral composition of the wing can be described using just three parameters: the anisotropy relaxation time τ, the internal friction (viscosity) relaxation time $\tau_\zeta = \zeta_0/(g_\infty - g_0)$, and the value of the modulus of elasticity g_0 at low frequencies.

The published experimental material on the wing of the Rayleigh line is quite extensive. A quantitative analysis of the near part of the wing and a qualitative analysis of the rest of the wing were provided by the early investigations [10, 31, 37–41]. The results of these investigations were compared with conclusions following from Leontovich's theory. It was established that Eq. (3) of this theory provides a correct description of many low-viscosity liquids in a wide range of temperatures. The physical nature of the wing and various features of the transport processes in liquid are discussed in these early papers.

The experimental results obtained by Starunov [8, 11, 12] and by Starunov and the present author [42] for low-viscosity liquids are in good agreement with the theoretical conclusions drawn in [8, 11, 12]. It was shown experimentally in [11, 12, 42] that the viscosity relaxation does indeed take place. Investigations of the spectral composition of the wing were used to find the values of τ_ζ for various low-viscosity liquids. Zaitsev and Starunov [42] confirmed experimentally that in the case of low-viscosity liquids τ_ζ is directly proportional to the viscosity, as required by the theory.

Thus, although the theoretical investigations of Leontovich, Rytov, and Starunov need further development, they provide a satisfactory description of the complex nature of the intensity distribution in the wing of the Rayleigh line and they explain much of what has not been understood before. However, many problems still remain. Some of them will be considered in the present paper.

§2. Width and Profile of Depolarized Lines of the Raman Scattering of Light in Liquids

Investigations of the width and profile of the Raman lines are of considerable interest because they can provide information on the rotational motion of molecules and on the intermolecular interaction in liquids. However, the width and profile of these lines are among the least known parameters of the Raman scattering spectra because of the considerable experimental difficulties which are encountered in their determination.

The width of the Raman lines was first considered theoretically by Sobel'man [43]. He showed that the width of the Raman lines may be affected by the random Brownian rotation of molecules.

The intensity of a Raman line is known to be proportional to the square of the amplitude E of the scattered wave, which itself depends on the derivative of the polarizability ε_{ik} with respect to the normal coordinate q_j, corresponding to a given molecular vibration. If the tensor $\varepsilon_{ik} = \partial \alpha_{ik}/\partial q_j$ is anisotropic, the amplitude E is modulated by the rotational motion of the molecules and it is a random function of time. If ε_{ik} is isotropic, the reorientation of molecules is not reflected in the value of E. In general, the tensor ε_{ik} consists of isotropic and anisotropic parts. The relationship between these parts governs the dependence of the width and profile of the Raman scattering line on the Brownian rotation. The same relationship determines also the integrated degree of depolarization of a line, Δ, which is a convenient criterion of the influence of the molecular reorientation on the width of various lines in the Raman spectrum.

We shall now consider three possible cases.

a) $\Delta = 6/7$; the trace of the tensor ε_{ik} vanishes. The incident light wave induces a molecular moment whose components can be found as in §1 [Eq. (9)]. The amplitude of the scattered wave, which is a random function of time, is governed by the correlation function $f(t)$. If the calculation is carried out within the framework of Starunov's theory [8, 11, 12], an expression is obtained for the intensity distribution in the profiles of the Raman lines and this expression is of the same form as Eq. (12) for the diffusion part of the wing.

b) $0 < \Delta < 6/7$; the tensor ε_{ik} has isotropic and anisotropic parts. This means that we can isolate the anisotropic part of the Raman scattering line, described by Eq. (12), and the isotropic part, whose width is independent of the molecular rotation.

c) $\Delta = 0$; the tensor ε_{ik} is isotropic and the Brownian rotation does not affect the line width. It must be stressed that the translational motion of the molecules can be ignored because it does not modulate the amplitude of the scattered light wave.

Valiev [44-46] showed that the width of the Raman lines can be affected also by the dissipative loss of vibrational quanta by a molecule (the conversion of the vibrational energy into thermal motion) and by the interaction of the molecule with its nearest neighbors. When these broadening factors are taken into account, the profile of a Raman line is given by the expression

$$I(\Omega) = I_0 \left[\frac{13}{6} \frac{\Delta}{1+\Delta} f(\gamma_1 + \gamma_2 + \gamma_3) + f(\gamma_1 + \gamma_2) \frac{6-7\Delta}{6(1+\Delta)} \right], \tag{22}$$

where

$$f(\gamma_1 + \gamma_2 + \gamma_3) = \frac{1}{\pi} \frac{\gamma_1 + \gamma_2 + \gamma_3}{(\gamma_1 + \gamma_2 + \gamma_3)^2 + \Omega^2},$$

$$f(\gamma_1 + \gamma_2) = \frac{1}{\pi} \frac{\gamma_1 + \gamma_2}{(\gamma_1 + \gamma_2)^2 + \Omega^2}.$$

Here, γ_1, γ_2, and γ_3 are the components of the line widths due to the interaction of the molecule with its neighbors, due to energy dissipation, and due to reorientation of molecules, respectively; I_0 is the peak value of the intensity of the Raman line; Ω is the frequency measured from the center of the line.

The formula (22) explains satisfactorily the experimentally determined widths of various Raman scattering lines. If $\Delta = 0$, the line width should be $\gamma_1 + \gamma_2$ and it should be independent of the molecular rotation. If the depolarization is $\Delta = 6/7$, the width is given by the sum $\gamma_1 + \gamma_2 + \gamma_3$, where the first two terms are practically independent of the temperature, whereas γ_3 (which represents the orientational motion) depends strongly on the temperature.

These and other features of the width of the Raman lines with different degrees of depolarization Δ were observed by Rakov [47-49], Sokolovskaya [50], and Rezaev [51]. Rakov investigated Raman lines with different degrees of depolarization and found the widths of these lines as a function of the temperature, viscosity, and the state of aggregation. He found that strongest temperature dependence was exhibited by the lines with $\Delta \approx 6/7$. The widths of the lines with small values of Δ was not greatly affected. Moreover, the width of the depolarized Raman lines could be represented by two parts which depended in different ways on the temperature: one of them was a function of the temperature and the other was independent of the temperature. In Valiev's theory the first width is γ_3, due to the rotational diffusion, and the second is the sum $\gamma_1 + \gamma_2$. When a substance goes over to the glassy state, in which the Brownian rotation is difficult or impossible, the widths of all the Raman lines become practically constant. Rakov called this the residual width. It is evident from the experimental results that the residual width $(\gamma_1 + \gamma_2)$ is different for different lines in the spectrum. If the residual width is subtracted from the total width of a depolarized line, we obtain the value of γ_3 which can be used to calculate the molecular reorientation time τ by means of Eq. (4).

Rakov made this calculation for the depolarized lines of cyclopentane, isopentane, and metaxylene and he found that the values of τ obtained in this way were is good agreement with the anisotropy relaxation times deduced from the diffusion part of the wing of the Rayleigh line. Moreover, a study of the temperature dependences of the viscosity η and the reorientation time τ, deduced from the widths of the depolarized Raman lines of these liquids indicated a direct proportionality between η and τ.

However, the published theoretical and experimental investigations are concerned solely with the near (up to 20 cm^{-1}) parts of the Raman line profiles. Studies of the far parts of these profiles should give additional information on the rotational motion of molecules in liquids. The faster types of rotational rocking of the molecules should be reflected in the far part of the wing of the depolarized lines. If this assumption is correct, the distribution of intensity in the far (exceeding 20 cm^{-1} from the center of the line) parts of these lines should be described by Eqs. (18) and (20), deduced by Starunov for the wing of the Rayleigh line.

In view of this it would be of great interest to compare the profiles of the depolarized Raman lines with the wing of the Rayleigh line in the widest possible range of frequencies. Such a comparison was carried out by Zaitsev and Starunov [20] and the results are given in Chap. III for six depolarized lines.

§3. Investigations of the Stimulated Scattering of Light in the Wing of the Rayleigh Line

In the preceding sections we considered the thermal wing of the Rayleigh line, which results from the modulation of light by fluctuations of the orientation of anisotropic molecules, which disperse with time. In the thermal scattering the weak field of the exciting light wave has a negligible influence on the anisotropy of the medium. However, when the scattering is excited by giant laser pulses, focused in a liquid, an allowance must be made for the influence of the light wave on the anisotropy of the medium. In this case the electric field of the exciting light wave is so high ($\sim 10^5$-10^7 V/cm) that, in combination with the thermal scattering field, it produces an orienting force [9] which acts on the anisotropic molecules in a liquid:

$$f \propto \frac{\alpha_1 - \alpha_2}{kT} E^2, \tag{23}$$

where α_1, $\alpha_2 = \alpha_3$ are the principal polarizabilities of the molecule in question; E is the sum of the fields of the exciting and scattered radiation. The low-frequency component of the above force orients the molecules and gives rise to an anisotropy which is the physical cause of the stimulated scattering of light in the wing of the Rayleigh line.

This nonlinear optical effect was discovered by Mash, Morozov, Starunov, and Fabelin-skii [16]. The fullest and most consistent theory of the stimulated wing of the Rayleigh line was developed by Starunov [8, 52].

Starunov considered symmetric-top molecules whose axes are oriented at angles ϑ_i and ϑ_k relative to a fixed system of coordinates (denoted by subscripts i and k). Starunov introduced an anisotropy tensor S_{ik} representing the deviation of the molecular axes from the isotropic distribution:

$$S_{ik} = \overline{\cos \vartheta_i \cos \vartheta_k} - \frac{1}{3} \delta_{ik}. \tag{24}$$

Let us assume that a plane light wave, polarized along the z axis so that $E_{z0} = E_0$ travels in a liquid of anisotropic molecules along the x axis. Then, the change in the anisotropy, resulting from the appearance of the force $f = f_{11} = \frac{4}{45} \frac{\alpha_1 - \alpha_2}{kT} E^2$, obeys the equation*

$$S + \tau \dot{S} = \frac{4}{45} \frac{\alpha_1 - \alpha_2}{kT} E^2. \tag{25}$$

Here, $S = S_{11} = \cos^2 \vartheta - \frac{1}{3}$ and $E = E_z = \frac{1}{2} \sum_{l=0}^{2} E_l(\mathbf{r}) \exp[i(\omega_l t - \mathbf{k}_l \mathbf{r})] +$ complex conjugate, where the subscripts 0, 1, 2 refer to the exciting, Stokes, and anti-Stokes waves, respectively. We note that Eq. (25) is valid only if the following condition is satisfied:

$$\frac{\alpha_1 - \alpha_2}{kT} E^2 \ll 1. \tag{26}$$

Starunov [8] deduced the following expression for S from Eq. (25):

$$S = \frac{2}{45} \frac{\alpha_1 - \alpha_2}{kT} \left\{ \sum_{l=0}^{2} |E_l|^2 + \left[\frac{e^{i\Omega t}}{1 + i\Omega t} (E_0 E_1^* \exp[-i(\mathbf{k}_0 - \mathbf{k}_1)\mathbf{r}] + \right. \right.$$
$$\left. \left. + E_0^* E_2 \exp[-i(\mathbf{k}_2 - \mathbf{k}_0)\mathbf{r}] + \text{complex conjugate} \right] \right\}, \tag{27}$$

where $\Omega = \omega_0 - \omega_1 = \omega_2 - \omega_0$.

The stimulated scattering of light in the wing of the Rayleigh line can be described by the simultaneous solution of Eq. (27) and of the Maxwell equations in which an allowance is

---

*The inertia of the orientational motion of molecules can be ignored at frequencies corresponding to the diffusion wing of the Rayleigh line. Therefore, the second derivative of S is ignored in Eq. (25).

made for the nonlinearity of the medium resulting from the extremely high intensity of the exciting light.

When **H** is eliminated from the Maxwell equations, we find that

$$\frac{\partial^2 \mathbf{D}}{\partial t^2} + c_0 \, \text{rot} \, \text{rot} \, \mathbf{E} = 0. \tag{28}$$

In a nonlinear medium the permittivity is a function of the field $\varepsilon(E) = \varepsilon_0 + \varepsilon_2 E^2$ or, in other words, the orientation of the molecules by the total electric field **E** gives rise to an additional permittivity $\Delta \varepsilon = \varepsilon_2 E^2$ and a corresponding nonlinear correction to the polarization P^{nl}. In this case, we obtain

$$\mathbf{D} = (\tilde{\varepsilon} + \Delta \varepsilon) \mathbf{E} = \mathbf{E} + 4\pi (\mathbf{P}^l + \mathbf{P}^{nl}), \tag{29}$$

where

$$\mathbf{P}^l = \frac{\tilde{\varepsilon} - 1}{4\pi} \mathbf{E}, \qquad \mathbf{P}^{nl} = \frac{\Delta \varepsilon}{4\pi} \mathbf{E} = \frac{1}{4\pi} \left(\frac{\partial \varepsilon}{\partial S} \right) S E. \tag{30}$$

The complex optical-frequency permittivity is $\tilde{\varepsilon} = \varepsilon' - i\varepsilon''$, where $\varepsilon' \approx n^2$, and $\varepsilon'' = 2n^2 k_\omega / |\,\mathbf{k}\,|$ Here, k_ω is the amplitude attenuation factor of light and **k** is the wave vector. When we substitute Eq. (29) into Eq. (28), we find that

$$\tilde{\varepsilon} \frac{\partial^2 \mathbf{E}}{\partial t^2} + c_0 \, \text{rot} \, \text{rot} \, \mathbf{E} + \frac{\partial^2}{\partial t^2} \left[\left(\frac{\partial \varepsilon}{\partial S} \right) S E \right] = 0. \tag{31}$$

Equations (27) and (31) form the system which we must now solve.

Starunov [52] solved the system for two cases, which are discussed in detail below.

1. If the Stokes and anti-Stokes waves do not interact with each other, the anti-Stokes wave decays rapidly and the exciting (E_0) and the Stokes (E_1) waves are found from Eqs. (27) and (31), neglecting the second derivatives of the amplitudes:

$$\left. \begin{array}{l} 2\mathbf{k}_0 \nabla E + 2 \dfrac{\omega_0}{c} k_\omega E_0 = - |\,\mathbf{k}_0\,|^2 \, A \, \dfrac{\Omega \tau}{1 + \Omega^2 \tau^2} \, |E_1|^2 \, E_0, \\[3mm] 2\mathbf{k}_1 \nabla E_1 + 2 \dfrac{\omega_1}{c} k_\omega E_1 = |\,\mathbf{k}_1\,|^2 \, A \, \dfrac{\Omega \tau}{1 + \Omega^2 \tau^2} \, |E_0|^2 \, E_1. \end{array} \right\} \tag{32}$$

In the system (32), we have

$$\left. \begin{array}{l} |\,\mathbf{k}_0\,|^2 = \dfrac{\omega_0^2}{c^2} \left\{ 1 + A \left[|E_0|^2 + |E_1|^2 \left(1 + \dfrac{1}{1 + \Omega^2 \tau^2} \right) \right] \right\}, \\[4mm] |\,\mathbf{k}_1\,|^2 = \dfrac{\omega_1^2}{c^2} \left\{ 1 + A \left[|E_1|^2 + |E_0|^2 \left(1 + \dfrac{1}{1 + \Omega^2 \tau^2} \right) \right] \right\}, \end{array} \right\} \tag{33}$$

where $A = \dfrac{2}{45} \dfrac{\alpha_1 - \alpha_2}{\varepsilon_0 kT} \dfrac{\partial \varepsilon}{\partial S} = \dfrac{\varepsilon_2}{\varepsilon_0}$ and $c = \dfrac{c_0}{\sqrt{\varepsilon_0}}$ (c_0 is the velocity of light in vacuum). If initially the amplitude of the exciting light wave is much higher than the amplitude of the Stokes wave, i.e., $|E_0| \gg |E_1|$, we find that when $|E_0|^2 \approx$ const the solution of the system (32) predicts an exponential rise of the intensity of the Stokes component in space (ξ is the coordinate in the direction of the wave vector $\mathbf{k}_1$ of the scattered light):

$$|E_1(\xi)|^2 = |E_1(0)|^2 \exp [g_1(\Omega) \xi], \tag{34}$$

provided the gain is $g_1(\Omega) > 0$.

In Eq. (34) the gain is

$$g_1(\Omega) = -2k_\omega + A|\mathbf{k}_1| \frac{\Omega\tau}{1 + \Omega^2\tau^2}|E_0(0)|^2. \tag{35}$$

The solution of the system (32) in the general case, when the amplitude of the Stokes wave $|E_1|$ need not be small for the scattering angle $\theta = 180°$ and $k_\omega = 0$, relates the initial values $|E_0(0)|^2$, $|E_1(L)|^2$ and the final values $|E_0(L)|^2$, $|E_1(0)|^2$ in the following way:

$$|E_0(L)|^2 = |E_0(0)|^2 - |E_1(0)|^2 + |E_1(L)|^2, \tag{36}$$

$$|E_1(L)|^2 = \frac{|E_1(0)|^2\left(1 - \frac{|E_1(0)|^2}{|E_0(0)|^2}\right)}{\exp\left\{|\mathbf{k}_0|A\frac{\Omega\tau}{1+\Omega^2\tau^2}(|E_0(0)|^2 - |E_1(0)|^2)L\right\} - \frac{|E_1(0)|^2}{|E_0(0)|^2}}. \tag{37}$$

It follows from the above relationship that during the initial stage of the development of a stimulated wing, when $|E_1| \ll |E_0|$, the intensity of the scattered light increases exponentially but when the amplitude of the Stokes wave reaches a value comparable with $|E_0|$, the saturation effect begins to be felt and the rate of rise of $|E_1|$ slows down.

An analysis of Eq. (35) shows that under some saturation conditions the highest value of the gain corresponds to a Stokes wave of the frequency $\omega = \omega_0 - 1/\tau$. This means that the stimulated scattering in the wing of the Rayleigh line can be excited most easily at the frequency corresponding to the half-width of the usual thermal wing [Eq. (4)] and the spectrum of the stimulated wing should have an intensity peak at this frequency. Thus, the position of the intensity peak can be used to find the anisotropy relaxation time τ.

Mash, Morozov, Starunov, and Fabelinskii [16] observed the scattered wing of the Rayleigh line in carbon disulfide, nitrobenzene, and salol. They found only the Stokes component of the wing and the intensity distributions had maxima at $\Delta\nu \sim 0.5$ cm^{-1} for nitrobenzene and $\Delta\nu \sim 0.16$ cm^{-1} for salol. Moreover, when the intensity of the exciting light was reduced slightly, the intensity of the Stokes component fell strongly, as required by the theory of the stimulated Rayleigh wing.

2. The foregoing theoretical analysis of the stimulated scattering of light in the wing of the Rayleigh line is limited to the case when the exciting, Stokes, and anti-Stokes waves do not interact with each other. However, it is predicted in [53] that this interaction should occur in the stimulated scattering through small angles. Starunov [52] considered this four-photon interaction in the linear approximation and he assumed that $|E_0|^2 \approx$ const, $|E_0| \gg |E_1|$, and $|E_0| \gg |E_2|$. In this case the Maxwell equations and Eq. (27) yield the system

$$\left.\begin{aligned}
2ik_1\nabla E_1 + \gamma_1 E_1 + 2i|\mathbf{k}_1|k_\omega E_1 &= |\mathbf{k}_1|^2 AE_0^2 E_2^* \frac{\exp[-i(2\mathbf{k}_0 - \mathbf{k}_1 - \mathbf{k}_2)\mathbf{r}]}{1 - i\Omega\tau}, \\
2ik_2\nabla E_2^* - \gamma_2 E_2^* + 2i|\mathbf{k}_2|k_\omega E_2^* &= -|\mathbf{k}_2|AE_0^{*2}E_1 \frac{\exp[-i(2\mathbf{k}_0 - \mathbf{k}_1 - \mathbf{k}_2)\mathbf{r}]}{1 - i\Omega\tau},
\end{aligned}\right\} \tag{38}$$

where

$$|\mathbf{k}_0|^2 = \frac{\omega_0^2}{c^2}(1 + A|E_0|^2). \tag{39}$$

Equation (38) has a new parameter $\gamma_l = \left\{k_l^2 - \frac{\omega_l^2}{c^2}\left[1 + A|E_0|^2\frac{1}{1-i\Omega\tau}\right]\right\}$ A strong interaction between the laser, Stokes, and anti-Stokes photons occurs only if

$$2\mathbf{k}_0 = \mathbf{k}_1 + \mathbf{k}_2. \tag{40}$$

We shall assume that the exciting radiation travels along the x axis. We shall also postulate that $k_{1y} = -k_{2y} = k_y$ and $k_{1x} \approx k_{2x} = k_x$. If we assume that the amplitude of the Stokes and anti-Stokes waves vary exponentially along the x axis, $E_1 \alpha \exp(\beta_1 x)$ and $E_2^* \alpha \exp(\beta_2 x)$, we obtain the following relationship which applies to small scattering angles provided the condition (40) is satisfied ($k_0 \approx k_x$):

$$\beta_1 = \beta_2 = \beta = -k_\omega \pm \frac{|k_0|}{2} \vartheta \left(\frac{2A|E_0|^2}{1 - i\Omega\tau} - \vartheta \right)^{1/2}, \tag{41}$$

where $\vartheta = k_y / k_0$. It follows from this expression that when

$$\vartheta_{\text{opt}}^2 = A|E_0|^2, \tag{42}$$

the maximum gain is the same for the Stokes and anti-Stokes parts of the wing:

$$g_2(\Omega) = -2k_\omega + \frac{|k_0| A |E_0|^2}{\sqrt{1 + \Omega^2 \tau^2}}. \tag{43}$$

It is interesting to note that the gain $g_2(\Omega)$ in the four-photon interaction differs from $g_1(\Omega)$ because it has a maximum at $\Omega = 0$, i.e., in this case the spectrum of the stimulated wing should include the Stokes and anti-Stokes components and the intensity maximum should coincide with the undisplaced frequency. Thus, the four-photon interaction in the stimulated scattering corresponding to the wing of the Rayleigh line should be observed experimentally in a narrow range of angles along a direction which is close to the direction of propagation of the laser radiation.

Experimental investigations of the four-photon interaction in the spectrum of the Rayleigh wing were carried out by the present author and the results will be described in Chap. IV.

Few experimental studies of the stimulated scattering in the wing of the Rayleigh line have been published so far. Apart from the work of Mash et al. [16], who discovered the stimulated wing, the study of Bloembergen and Lallemand [54] is of great interest.

Bloembergen and Lallemand used a Q-switched ruby laser emitting two modes separated by a frequency interval 1.6 cm^{-1}. The ruby laser radiation was focused in a cell containing cyclohexane. The width of the stimulated Raman scattering lines was investigated. Different amounts of carbon disulfide, a liquid consisting of anisotropic molecules, was then added to the cell. The Raman lines of cyclohexane increased in width with increasing concentration of carbon disulfide, reaching 85 cm^{-1} for a volume concentration of about 15% CS$_2$.

Bloembergen and Lallemand explained these results as follows. When two light waves of different frequencies ω_1 and ω_2 reached the cell containing the solution of carbon disulfide in cyclohexane, the total electric field of frequency $(\omega_2 - \omega_1)$ began to orient the anisotropic molecules of carbon disulfide and this gave rise to a periodic variation in the refractive index of the solution. When a third wave of frequency ω_3 different from ω_1 and ω_2 (in this experiment the latter was the Raman line of cyclohexane), reached the cell, this wave became modulated and broadened. Thus the mechanism was the same as in the stimulated scattering in the Rayleigh wing.

Rank and his colleagues [55] carried out a study of the influence of temperature on the stimulated wing of the Rayleigh line of nitrobenzene and meta-nitrotoluene. They used a Q-switched ruby laser for the excitation of the stimulated scattering. They found that the spectrum of the light scattered in a nitrobenzene included a line whose position shifted from 0.09 to 0.41 cm^{-1} (measured from the laser line) when the temperature was raised from 285 to $390°$K. Rank et al. made use of the results given in [16] and interpreted this line as the degenerate stimulated Rayleigh wing.

However, in our opinion the stimulated wing cannot degenerate into a line because many Stokes waves with the gain $g_1(\Omega)$ given by Eq. (35) will be amplified when the laser radiation power exceeds the threshold value. It is more likely that the line observed by Rank et al. [55] is a weak laser mode which is amplified in the liquid containing anisotropic molecules by the same mechanism as the Stokes component of the wing. The temperature dependence of the separation between this mode and the principal laser mode is a natural consequence of the temperature dependence of the gain, which is related to the relaxation time τ [Eqs. (3) and (35)]. We shall return to this point later.

Thus, detailed investigations of the stimulated scattering in the wing of the Rayleigh line make it possible to understand more fully the nature of the interaction of light with matter.

CHAPTER II

APPARATUS AND METHODS USED IN INVESTIGATIONS OF THE SPECTRA OF THE THERMAL AND STIMULATED SCATTERING OF LIGHT

§1. Method Used in Investigations of the Thermal Rayleigh Wing and of the Raman Line Profiles

The thermal wing of the Rayleigh line and the profiles of the depolarized Raman lines of liquids were investigated using specially assembled apparatus with photoelectric recording. This type of recording made it possible to obtain directly the frequency distribution of the intensities. Moreover, the use of a photomultiplier in place of a photographic film eliminated the photometric errors. Finally, photoelectric recording of the intensity yielded the results much faster.

We used a DFS-12 spectrograph with two replica diffraction gratings of 140×150 mm size and with 600 lines/mm. The gratings were used in the second diffraction order, in which the linear dispersion was 4.6 Å/mm. When the entry and exit slits were ~ 2-4 μ wide, the half-width of the transfer function (including the natural width of the 4358 Å mercury line) was 0.6 cm^{-1}. This enabled us to study the part of the wing adjoining the undisplaced line.

Figure 1 is a schematic diagram of the apparatus. The light emitted by two low-pressure mercury lamps Q [56] was directed by aperture-limiting diaphragms A and by cylindrical lenses into a cuvette V containing the liquid under investigation. The cuvette was constructed in such a way that only the light scattered by the liquid but not the light reflected from the cuvette walls reached the spectrograph [10]. Diaphragms D_1 and D_2 helped in achieving this situation. The scattered light passed through a mechanical modulator O (which interrupted the light flux at a frequency of 415 Hz), a condenser L_1, a polarizer N, and the entry slit S_1 before it reached the diffraction grating. The scattered beam followed a complicated path within the spectrograph (shown by arrows in Fig. 1). The part of the scattered-light spectrum selected by the exit slit S_3 was directed by a cylindrical lens L_2 onto the cathode of a photomultiplier PM. The photomultiplier (EMI 6094B) had a high signal-to-noise ratio in the blue part of the spectrum. The electric signal produced by the photomultiplier passed through a preamplifier Y_1, an amplifier Y_2, and a phase-sensitive detector to an ÉPP-09 electronic potentiometer.

The reference voltage required in phase detection was provided by a photoresistor F, which was illuminated through the modulator O by an additional source of light S (an incandescent lamp supplied by a dc source).

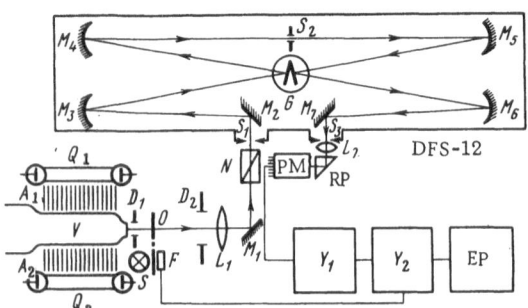

Fig. 1. Schematic diagram of the photoelectric apparatus used in investigations of the thermal wing of the Rayleigh line. Q_1 and Q_2 are low-pressure mercury lamps; A_1 and A_2 are aperture-limiting diaphragms; V is a cuvette containing the scattering liquid; D_1 and D_2 are diaphragms; O is a mechanical modulator of the scattered light (415 Hz); S is an additional source of light; F is a photoresistor providing a reference voltage for synchronous detector control; L_1 and L_2 are lenses; N is a polarizing prism; M_1-M_7 are mirrors; S_1, S_2, and S are spectrograph slits; G is a pair of replica gratings with 600 lines/mm; PM is a photomultiplier; RP is a rotatable prism; EP is an electronic potentiometer; Y_1 and Y_2 are amplifiers.

The replacement of the standard circuit with a dc amplifier by an ac amplifier and a phase-sensitive detector had many advantages. First, the gain achieved in this way was higher; secondly, this arrangement eliminated the "zero drift," which is a characteristic of dc amplifiers. This improved the stability and reliability of the recording unit. A careful check was made of the linearity of the response of the electronic circuit throughout the range of conditions used in our investigation.

Before we could investigate the near part of the wing, we had to eliminate the isotropic (z-polarized) component of the scattered light. The x component of the scattered-light spectrum up to 50 cm^{-1} was selected by the polarizing prism N. The far part of the wing was recorded without the polarizer, which enabled us to extend measurements up to 130 cm^{-1}.

The scattering region was illuminated with natural (unpolarized) light in all the investigations except in the study of the angular distribution of the intensity in the wing of the Rayleigh line. In this investigation a Polaroid was used to select either the x or the z component of the exciting radiation but this reduced the intensity of the scattered light. Nevertheless, the spectrum of the wing could be studied up to 90-100 cm^{-1} measured from the undisplaced frequency.

The angle between the incident and the scattered beams was measured with a special louver placed between the mercury lamps and the cuvette containing the scattering liquid.

In the investigation of the intensity distribution in the spectrum of the depolarized Raman lines, which were weaker than the Rayleigh lines, we used a two-lamp illuminator with an elliptic reflector.

An examination of the operation of the DFS-12 spectrograph showed that the replica gratings produced weak "ghosts" at frequencies of 6.7, 12.5, and 25 cm^{-1}. However, the intensity of these "ghosts" did not exceed 1% of the intensity at the peak of the exciting line. In the investigation of the profile of the Rayleigh wing of low-viscosity liquids the intensity of the "ghosts" did not exceed the noise level of the recording part of the apparatus. However, when the same wing was studied in viscous liquids, it was found that the "ghosts" were significant because most of the scattered light was concentrated in a narrow region adjoining the undisplaced frequency. It was then impossible to study the near part (up to 30 cm^{-1}) of the wing.

The distortion of the profile of the Rayleigh wing by the "ghosts" was determined by recording in all cases not only the spectrum of the wing but also the exciting line at an intensity 1.5-2 times higher than the intensity of the scattered light.

The results reported in Chap. III were obtained by averaging three to five consecutively recorded spectra. The scatter of the points in the central part of the wing (up to 60 cm^{-1}) did not exceed 5%. The random errors in the other parts of the spectrum are indicated by vertical lines in the relevant figures.

It was convenient to analyze the intensity distribution in the Rayleigh wing and in the wings of the Raman lines by plotting the experimental data using the coordinates $1/I$ (the reciprocal of the intensity) and $\Delta\nu^2$ (the square of the frequency measured from the center of the line).

§2. Apparatus Used in Investigations of the Stimulated Scattering of Light

Figure 2 is a schematic diagram of the apparatus used in the excitation and observation of the stimulated scattering in the wing of the Rayleigh line and of the stimulated Brillouin scattering.

A Q-switched ruby laser was used as the excitation source. Our laser consisted of two ruby crystals (R_1 and R_2), which were 120 mm long and 10-12 mm in diameter. The concentration of the Cr^{3+} ions in these crystals was 0.05%. The optic axes of the crystals were approximately normal to the axes of the rods. The lateral surfaces of the rods were either polished or mat. Each of the rods was placed, together with two IFP-2000 flashlamps, in an elliptic reflecting enclosure which was cooled by circulating tap water. Each lamp was subjected to a voltage of 1200-1500 V supplied by a capacitor bank. The optic axes of the ruby crystals were parallel. The laser radiation was polarized along the vertical (z) axis. An optical resonator, enclosing the ruby rods R_1 and R_2, was formed by a mirror M_1 with a dielectric coating (reflection coefficient ~ 100% at $\lambda = 6943$ Å) and two plane-parallel plates m_1 and m_2 with a common reflection coefficient of ~ 15%. The Q factor of the resonator was modulated by two cells C_1 and C_2 which had plane-parallel windows and were filled with a solution of cryptocyanine in ethyl alcohol. One cell (C_1) was placed between the mirror M_1 and the first ruby R_1; the second cell C_2 was located between the two ruby rods. The thickness of each cell was 10 mm and the transmission coefficient at $\lambda = 6943$ Å was ~ 8% for C_1 and ~ 2% for C_2.

The plates m_1 and m_2 were used also as the mode selector. These plates acted as a complex multiple-reflection interferometer because their thicknesses differed quite considerably (the plate m_1 was 2.5 mm thick and the plate m_2 was 12.6 mm). Inside the laser there were many reflecting surfaces which formed resonators with different spectral ranges. Consequently, several modes could be emitted by the laser. The multiple-reflection interferometer selected that mode which coincided with the natural modes of the resonators m_1 and m_2. The degree of coincidence of the modes of the resonators m_1 and m_2 could be varied by inclining one of the plates. Usually satisfactory mode selection was achieved by inclining slightly the ruby rod R_2 in the vertical plane (in the case of z-polarized radiation) relative to the ruby R_1. The spectral width of the line emitted by the laser was less than 0.05 cm^{-1}. Only the lowest-frequency modes of the resonators formed by the surfaces M_1 and m_2 or M_1 and m_1 could fit within this width. When the pumping level was slightly higher than the threshold value, the laser produced one pulse for every flash emitted by the lamps.

The pulse duration, measured at mid-amplitude, was 10-15 nsec. The power of the giant pulses emitted by the laser was of the order of 100 MW.

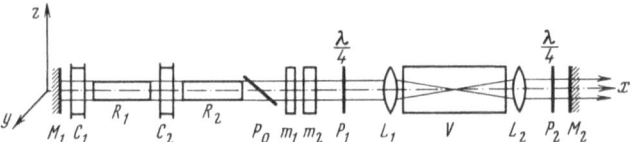

Fig. 2. Schematic diagram of the apparatus used
for the excitation of the stimulated scattering in
the wing of the Rayleigh line. M_1 and M_2 are mirrors with the dielectric coatings ($r_1 \sim 100\%$, $r_2 \sim$
60%); C_1 and C_2 are cuvettes filled with solutions
of cryptocyanine in ethyl alcohol; R_1 and R_2 are
ruby rods; m_1 and m_2 are plane-parallel plates
used as mode selectors; P_0 is a glass plate aligned
at the Brewster angle; P_1 and P_2 are quarter-wave
plates; L_1 and L_2 are lenses forming a telescopic
system; V is a cuvette containing the scattering
liquid.

The stimulated scattering of light was observed when the ruby laser radiation was focused by a lens L_1 inside the cuvette V containing the liquid under investigation. The forward-
scattered light and the laser radiation passed on into the optical part of the apparatus. The
light scattered through 180° returned back to the laser, was amplified there, and it again
reached the cuvette V. The mixture of light scattered at angles of 0° and 180° was directed,
via a light filter, to a Fabry-Perot interferometer. In the investigation of the four-photon interaction the spectrum of the stimulated wing was recorded simultaneously at scattering angles
of 0° and 90°. Next, the spectroscopically analyzed light was focused by an objective of $f =$
1200 mm onto a photographic film. The spectral ranges of the interferometers were 5 and
8.33 cm^{-1} in the study of the stimulated scattering in the wing of the Rayleigh line and 1 cm^{-1}
in the study of the stimulated Brillouin scattering.

In some cases the forward-scattered light had to be amplified in the laser without admis-
sion of the back-scattered light. In such cases two quarter-wave plates P_1 and P_2 were located
in front of the cuvette V and behind it. A glass plate P_0, inclined at the Brewster angle, was
inserted in the resonator (Fig. 2). The lens L_1 was supplemented by a second lens L_2 in such
a way that the lens system became telescopic. Usually the focal lengths of the two lenses were
equal. All these elements were located inside the external resonator formed by the plates m_1
and m_2 and a mirror with a dielectric coating whose reflection coefficient was $\sim 60\%$ at $\lambda = 6943$ Å.
The linearly polarized (along the z axis) ruby laser radiation was converted by P_1 to the circular polarization. The back-scattered light also had the circular polarization, which was converted by P_1 to the linear polarization directed along the y axis. Such scattered light could not
be amplified in the ruby laser. The forward-scattered light, which had the circular polarization, passed through the second quarter-wave plate P_2 and was converted into light polarized
along the y axis. This was followed by partial reflection from the semitransparent mirror M_2
(reflection coefficient $r_2 \sim 60\%$), a second pass through the plate P_2, the cuvette, and the plate
P_1. In this way, the forward-scattered light, which was reflected from the mirror M_2 and which
passed through the two quarter-wave plates, acquired the polarization directed along the z axis.
Consequently, such light could be amplified in the laser.

In the arrangement described above the entry and exit windows of the cuvette had to satisfy quite stringent requirements. They had to be plane and mutually parallel. We used specially prepared glass cuvettes with carefully polished end flanges. The windows were then attached to the flanges in such a way as to ensure good optical contact.

The scattered-light spectra were investigated in the following sequence. First, we photographed the spectrum of the ruby laser radiation with the aid of an interferometer. When the single-mode emission was achieved, the output energy of the laser was measured. Then, the stimulated scattering spectra were photographed. Finally, the spectrum of the laser radiation was recorded again.

The interferograms of the scattered-light spectra were interpreted by the off-center method [57]. In studies of the stimulated wing the interferograms were analyzed photometrically with an MF-4 microphotometer.

CHAPTER III

RESULTS OF INVESTIGATIONS OF THE SPECTRUM
OF LIGHT SCATTERED BY ANISOTROPY FLUCTUATIONS
IN LIQUIDS

§1. Wing of the Rayleigh Line of Viscous Liquids

The spectral distribution of the intensity of light scattered in three viscous liquids was investigated at various temperatures. In such investigations the liquids have to satisfy several conditions. The viscosity of the liquid should vary strongly with the temperature but the liquids should not crystallize at high values of the viscosity. Moreover, the wing of the Rayleigh line should be quite strong and wide and the liquid should consist of anisotropic molecules.

These requirements are satisfied by benzophenone, salol (phenyl salicylate), and triacetin. The benzophenone and salol molecules are strongly anisotropic, as indicated by the depolarization coefficient Δ of the scattered light ($\Delta = 0.75$ for benzophenone and salol [1]). However, benzophenone and salol are in crystalline state at room temperature. Therefore, these liquids must be studied in the supercooled state, which is attained by repeated heating to the boiling point and rapid cooling. After this treatment salol and benzophenone remain liquid for a long time below the crystallization point. The situation in respect of triacetin is somewhat simpler because this liquid does not crystallize on transition to the solid state. However, the triacetin molecules are less anisotropic than the molecules of salol and benzophenone: the depolarization coefficient of the light scattered in triacetin is $\Delta = 0.38$ [58].

The intensity distribution in the wing of the Rayleigh line of salol was investigated at temperatures of $-38, 0, 23$, and $87°C$; triacetin was studied at $-20, 23$, and $51°C$ and benzophenone at $-50, 23$, and $82°C$. The viscosity of these liquids varied by several orders of magnitude in the range of temperatures investigated. For example, Velichkina [58] found that the viscosity of triacetin at $-20°C$ was 15 P but at $51°C$ the viscosity was only 0.021 P. The wing of the Rayleigh line of the three liquids was investigated only in the range 30-130 cm^{-1}, measured from the frequency of the exciting radiation. The region closer to the undisplaced line could not be studied because the diffraction "ghosts" were too strong.

Experiments showed that the frequency distribution of the intensity in the Rayleigh wings of the three viscous liquids was similar to the frequency distribution of the intensity of the light scattered by low-viscosity liquids. In both cases the wing could be divided into three regions where the intensity varied in different ways. In the first region the intensity fell very rapidly with increasing frequency; this was followed by a slower fall of the intensity and, by a region where the intensity again decreased rapidly. It was interesting to note that in the third region the intensity varied with the frequency must faster than in the first region. The second

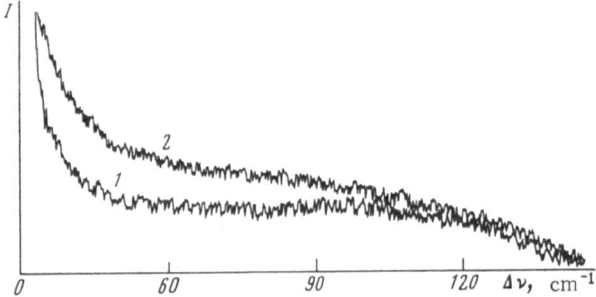

Fig. 3. Thermal wing of the Rayleigh line of
benzophenone at $-50°C$ (1) and $23°C$ (2).

region, where the intensity decreased slowly, expanded and became nonmonotonic when the viscosity was increased (Fig. 3). This nonmonotonic variation in the wing of the light scattered
in liquid benzophenone was reported by Venkateswaran [28], who attributed this variation to
the valence deformation vibrations of the two benzene rings of the benzophenone molecule. Our
interpretation of the nonmonotonic variation in the wing will be given later.

There is as yet no molecular theory of the Rayleigh wing of viscous liquids because there
is no general theory of high-viscosity liquids. However, the phenomenological theory of the
Rayleigh wing developed by Rytov [4] applies to liquids of any viscosity. Therefore, it would
be interesting to analyze the experimental data on the basis of this theory. However, we must
first determine the meaning of the function $X(\Omega)$, which occurs in Eq. (8).

We shall assume that the permittivity tensor ε_{ik} is proportional [3] to the anisotropy
tensor S_{ik}. The time dependence of S_{ik} is given by the reaction equation

$$b\ddot{S}_{ik} + \xi\dot{S}_{ik} + gS_{ik} = \eta\left(\dot{e}_{ik} - \frac{1}{3}e_{ii}\delta_{ik}\right),\tag{44}$$

or, in the spectral approach by the equation

$$S_{ik}\left(-b\Omega^2 + i\Omega\zeta + g\right) = i\Omega\eta\left(e_{ik} - \frac{1}{3}e_{ii}\delta_{ik}\right).\tag{45}$$

In the above expressions b is a parameter associated with the inertia; ζ represents the friction
experienced in the dispersal of the anisotropy fluctuations; g is the elasticity of the medium in
the case of sudden departures from isotropy. The last two quantities (ζ and g) represent the
reaction of the medium exerted on the molecules during their orientational motion and they
should vary with frequency in accordance with the same law as the viscosity η and the shear
modulus μ which represent the reaction of the medium exerted on the molecules during their
translational motion (shear deformation).* The dispersion laws of ζ and g may be assumed to
be given by the expressions in Eq. (20). The term on the right-hand side of Eq. (45) should now
be replaced with

$$\eta_0\frac{i\Omega}{1+i\Omega\tau_\eta}\left(e_{ik} - \frac{1}{3}e_{ii}\delta_{ik}\right),$$

which corresponds to a viscous force at low frequencies and an elastic force at high frequencies. In this way we obtain the expression

$$S_{ik}\left[-b\Omega^2 + i\Omega\zeta(\Omega) + g(\Omega)\right] = \eta_0\frac{i\Omega}{1+i\Omega\tau_\eta}\left(e_{ik} - \frac{1}{3}e_{ii}\delta_{ik}\right),\tag{46}$$

and hence we can find the function $X(\Omega)$:

$$|X(\Omega)|^2 = \frac{\eta\eta_0}{g}\frac{\Omega^2}{\left(1 - \frac{b}{g}\Omega^2\right)^2 + \Omega^2\left(\frac{\zeta}{g}\right)^2}.\tag{47}$$

_____

* The difference between the dispersion laws of these quantities can affect only the relaxation
times of the internal friction τ_ζ and of the viscosity τ_η.

If we substitute Eq. (47) into Eq. (8) we obtain the required expression for the intensity distribution in the wing of the Rayleigh line,

$$I(\Omega) = \text{const}\, \frac{kT}{8\pi} \frac{\eta(\Omega^2)}{g^2} \frac{1}{(1 - a\Omega^2)^2 + \Omega^2\tau^2}, \tag{48}$$

where $a = b/g$ and $\tau = \zeta/g$. It must be stressed that ζ and g depend on the frequency Ω and, consequently, a and τ are — generally speaking — functions of Ω.

An allowance for the inertial term and for the relaxation of the viscosity, internal friction, and elasticity means that Eq. (48) deduced from the phenomenological theory of relaxation takes account of the most important of the processes affecting the spectral composition of the scattered light.

The intensity distribution in the wing can be calculated theoretically by means of Eq. (48) and it can be compared with the observed distribution provided we know a and τ for each liquid. Unfortunately, no information is available on these two parameters and, therefore, it is necessary to estimate their values using Eq. (48) and the experimentally determined intensity distribution in the wing. However, a comparison can still be carried out by matching just one point of the theoretical dependence of Eq. (48) to the experimental curve. In this way we can determine to what degree the theory agrees with the experimental results at other points in the spectrum.

The values of a and τ can be found in any spectral range by taking the values of the intensities at three frequencies in a narrow spectral range (for example, in the high-frequency part of the wing) and substituting these values in Eq. (48). The results of such an analysis of spectrograms are given in Table 1. It is clear from this table that — in the investigated temperature range and at frequencies corresponding to the far part of the wing — the parameters a and τ vary slowly with the temperature. This means that the quantities associated with a and τ, namely the limiting value of the rotational rocking frequency $\Omega_c = (g_\infty/b)^{1/2}$ and the high-frequency internal friction ζ, also vary weakly with the temperature.

It is worth mentioning that the limiting frequency of the rocking oscillations of the salol and benzophenone molecules in liquids are in good agreement with the low-frequency spectra of the corresponding molecular crystals. According to Venkateswaran [28], the low frequencies in the benzophenone spectrum are $\nu_1 = 102$ cm^{-1}, $\nu_2 = 122$ cm^{-1} and the corresponding frequency in the spectrum of salol is $\nu = 96$ cm^{-1}. It follows that at the frequencies corresponding to the far part of the wing the modulus of elasticity g is close to its limiting value g_∞ and the liquid behaves as a solid.

TABLE 1

Substance	t, °C	$a \cdot 10^{23}$, sec^2	Ω_c, cm^{-1}	$\tau \cdot 10^{14}$, sec
Salol	−33	3,0	98	3,2
	0	3,1	96	3,2
	23	3,0	97	2,9
	87	3,1	96	3,2
Benzophenone	−50	2,2	114	2,4
	23	2,3	111	2,4
	82	2,3	111	2,2
Triacetin	−20	5,4	73	6,6
	23	4,2	82	6,2
	51	5,4	73	6,4

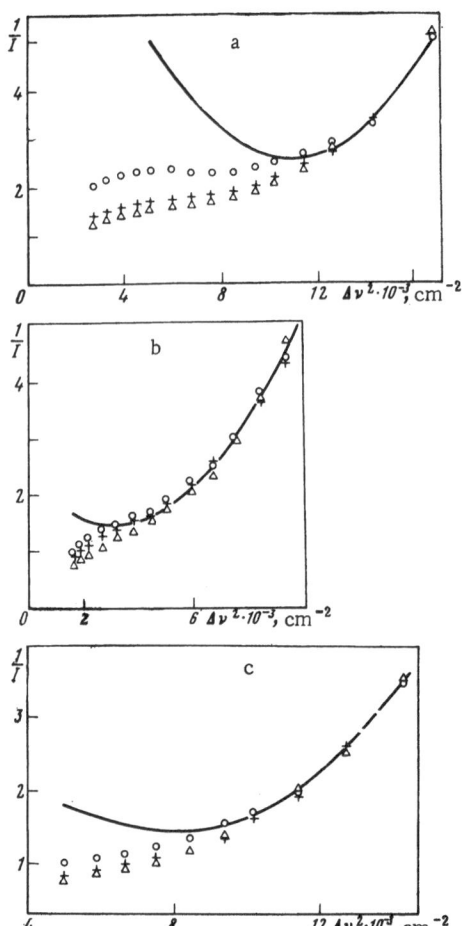

Fig. 4. Dependences of the reciprocal of the intensity in the wing of the Rayleigh line on the square of the frequency, plotted for triacetin (a), benzophenone (b), and salol (c). The continuour curves are theoretical dependences described by Eq. (48). The circles, crosses, and triangles are the experimental values of the reciprocal of the intensity of light scattered at −20, 23, and 51°C (a), −50, 23, and 82°C (b), and −38, 23, and 87°C (c).

The high-frequency internal friction in viscous liquids is independent of the temperature because it varies in accordance with the law

$$\zeta = \frac{\zeta_0}{1 + \Omega^2 \tau_\zeta^2} + \zeta',$$

$$(49)$$

where ζ' is some component of the internal friction which is independent or weakly dependent on the temperature and frequency. It follows from this formula that at high frequencies and for large values of the relaxation time of the internal friction τ_ζ (which is proportional to the viscosity), the first term in Eq. (49) is very small and therefore the internal friction is dominated by ζ'. The experimental values of ζ' deduced from the above equation are 2.6×10^{-24} and 1.9×10^{-24} g·cm²·sec⁻¹ for salol and benzophenone, respectively.

The nonmonotonic variation of the intensity of the depolarized scattered light at low temperatures is due to the relaxation of the internal friction and the "rigidity" of liquids at high frequencies.

The rate of frequencies at which the relaxation of the internal friction (viscosity) became important was found by comparing the experimental results with the theoretical dependence of Eq. (48), which was plotted using the values of a and τ deduced from the far parts of the wing. It was assumed that these parameters did not vary with frequency, i.e., that the internal friction was practically independent of Ω.

The results of this comparison are plotted in Figs. 4a, 4b, and 4c for triacetin, benzophenone, and salol, respectively. The ordinates in Fig. 4 are reciprocals of the intensity in the wing, 1/I, and the abscissas represent the square of the frequency, $\Delta\nu^2$, measured from

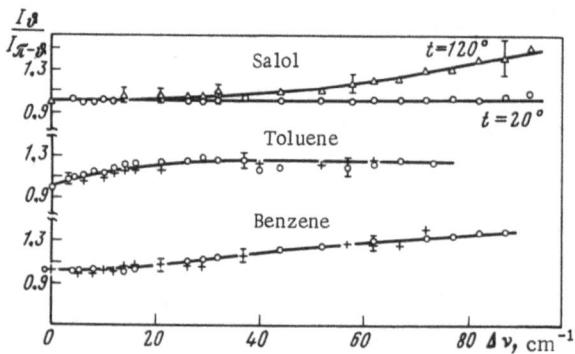

Fig. 5. Frequency dependence of the ratio $z_1(\Omega)$ for the wing of the Rayleigh line. The circles represent the z-polarized exciting light and the crosses the x-polarized light.

the center of the undisplaced line. The continuous curves in Figs. 4a–4c show the theoretical distribution of the intensity corresponding to Eq. (48) and calculated on the assumption that the values of a and τ are constant. The circles, crosses, and triangles are the experimental values of $1/I$ obtained at different temperatures. The theoretical curves and the experimental results were made to fit* at 87 cm^{-1} in the case of triacetin and at 120 cm^{-1} in the case of benzophenone and salol.

The experimental and the calculated results agreed in the range from 60 to 100 cm^{-1} in the case of triacetin, from 110 to 130 cm^{-1} for benzophenone, and from 100 to 120 cm^{-1} for salol. In all cases the range of frequencies in which agreement was obtained increased with decreasing temperature of the liquid. The limited agreement indicated that the relaxation of the internal friction was completed at the frequency corresponding to the lower limit of the interval of coincidence between theory and experiment.

The disagreement between the experimental data and the calculations based on Eq. (48) and Table 1, which was observed at frequencies below 60, 110, and 100 cm^{-1} for triacetin, benzophenone, and salol, respectively, suggested that the first Ω-dependent term im Eq. (49) for ζ became important in this range of frequencies. However, it was not possible to make an allowance for the influence of this relaxation term because of the lack of any data on the relaxation time of the internal friction τ_ζ in the liquids under investigation.

The theory predicts that in the case of high values of τ_ζ the spectrum of the depolarized scattered light should have a maximum at the frequency

$$\Omega_{max} = \left(\Omega_c^2 - \frac{\zeta^2}{2b} \right)^{1/2}. \tag{50}$$

However, the experimental intensity maxima, corresponding to the minima of the dependences $1/I = f(\Delta\nu^2)$, were closer to the undisplaced line than the maxima predicted theoretically (Fig. 5). This could be explained easily by the presence of a constant component of the internal friction ζ' in a viscous liquid because an increase in the second term in Eq. (50) would shift the intensity maximum in the direction of lower frequencies.

Thus, the phenomenological theory of the scattering of light [4] can explain the influence of temperature on the wing of the Rayleigh line of viscous liquids.

---

* The experimental and the theoretical curves were made to fit at a point which was selected arbitrarily. Therefore, the positions of the other points yielded information on the nature of the frequency dependence of the intensity at various temperatures but not whether the absolute intensity of the scattered light increased or decreased (in a given frequency range) when the temperature was altered.

§2. Angular Distribution of the Intensity in the
Thermal Wing of the Rayleigh Line

The angular distribution of the intensity of the scattered light is a valuable source of information on the physical properties and features of the scattering medium. This has provided the stimulus for recent investigations of this distribution [59-62]. However, these investigations were concerned either with the angular distribution of the integrated intensity of the Rayleigh scattering [59] or of the indicatrix of the Raman scattering lines [60-62]. It was found that the indicatrix was asymmetrical with the predominant forward scattering.

A theoretical analysis [1, 63] of the angular distribution of the integrated intensity of light scattered by that component of the anisotropy fluctuations which is described by the symmetric anisotropy tensor leads to the expression

$$I \propto \left(1 - \frac{\sin^2 \theta}{14}\right), \tag{51}$$

if the scattering is excited by natural (unpolarized) light, and to

$$I \propto \left(1 + \frac{\sin^2 \vartheta}{6}\right), \tag{52}$$

if the incident light is polarized. In Eqs. (51) and (52), θ is the scattering angle and ϑ is the angle between the electric field $\mathbf{E}$ in the incident light wave and the direction of observation of the scattered light. The angular dependence of the integrated intensity of the light scattered by the antisymmetric part of the anisotropy tensor is given by the formulas:

$$I \propto \left(1 + \frac{\sin^2 \theta}{2}\right), \tag{53}$$

$$I \propto (1 + \cos^2 \vartheta), \tag{54}$$

which apply to the natural and polarized incident light, respectively.

It should be mentioned that Eqs. (51)–(54) are derived on the assumption that the anisotropy fluctuations are statistically independent in two neighboring regions in the scattering medium.

Cohen and Eisenberg [59] observed a symmetric angular distribution of the integrated intensity, which was in good agreement with the formulas given above.

It seemed highly desirable to investigate the angular dependence of the spectral composition in the wing of the Rayleigh line.

The scattering spectra were determined in the wing of the Rayleigh line for benzene, carbon disulfide, toluene, carbon tetrachloride, salol, and benzophenone. This determination was carried out in the frequency range 0-100 cm^{-1} for three scattering angles of 60, 90, and 120°.

The molecules in these liquids differed in their shape, dimensions, inertia, and dipole moments. The liquids selected included those of low viscosity (benzene, carbon disulfide, toluene, and carbon tetrachloride) and high viscosity (salol and benzophenone).

The measurements were carried out as follows. The fluorescence of a weak solution of quinine sulfate (concentration 5×10^{-5} g/ml [64]) was first recorded at 60°. This was done because the indicatrix of the luminescence of quinine sulfate was known to have spherical symmetry. Next, keeping the conditions and the scattering angle constant, we recorded the wing of the Rayleigh line and then the fluorescence spectrum for a second time. This was repeated for the angles of 90 and 120°.

An analysis of the spectrograms for each of the investigated liquids yielded the dependences of the ratios

$$z_1(\Omega) = \frac{I_{60}(\Omega)}{I_{120}(\Omega)},$$

$$z_2(\Omega) = \frac{I_{90}(\Omega)}{I_{120}(\Omega)}$$

on the frequency of the scattered light. Next, $z_1(\Omega)$ and $z_2(\Omega)$ were multiplied, respectively, by the ratio $I_{quin}(120°)/I_{quin}(60°)$ and $I_{quin}(120°)/I_{quin}(90°)$ in order to eliminate a possible difference between the scattering volumes corresponding to the angles in question and the difference between the intensities of the exciting radiation. The error [65] in the determination of $z_1(\Omega)$ and $z_2(\Omega)$ due to the difference between the refractive indices of the solution of quinine sulfate and of the investigated liquids was less than 1% and it was ignored.

Benzene, toluene, and carbon disulfide were illuminated with the x- and z-polarized light. Carbon tetrachloride, salol, and benzophenone were illuminated with natural light. In all cases the x component was isolated in the scattered-light spectrum up to 40 cm^{-1} from the undisplaced line.

The experimental results obtained for some of the liquids are plotted in Fig. 5. It is evident from this figure that the ratio $z_1(\Omega)$ for benzene is unity in the 0.20 cm^{-1} range, whereas at higher frequencies (20-90 cm^{-1}) the intensity of the forward-scattered light exceeds the intensity of the back-scattered light. A similar dependence is observed for $z_1(\Omega)$ of toluene at room temperature and for salol at 120°C. In the case of carbon disulfide, carbon tetrachloride, benzophenone, and salol the indicatrix is symmetric at room temperature throughout the investigated frequency range.

The ratio $z_2(\Omega)$ is 0.81 for all the liquids if the exciting radiation is polarized and 0.98 if the exciting radiation is unpolarized. The values of $z_2(\Omega)$ are in good agreement with the conclusions that follow from the theory of scattering as a result of those anisotropy fluctuations which can be described by the symmetric anisotropy tensor.

It is important to note that the forward scattering becomes significant in benzene, toluene, and salol at t = 120°C, beginning from some particular frequency.

The results obtained cannot be explained by means of any of the theories of scattering which ignore the correlation between the anisotropy fluctuations in two neighboring regions of the scattering medium.

Let us now consider the influence of the mutual correlation in the orientational motion of neighboring molecules in a liquid on the angular distribution of the scattered light.

Let us assume that a light wave polarized in the horizontal plane travels in a liquid consisting of anisotropic molecules which are of the symmetric-top type (principal polarizabilities α_1, $\alpha_2 = \alpha_3$). We shall specify the orientation of each molecule by the angular coordinates ϑ and φ. We shall assume that the scattered light is observed at an angle θ with respect to the direction of the wave vector of the incident wave. This incident wave induces, in the i-th molecule, a moment whose components are

$$\left. \begin{aligned} P_y^i &= E(\alpha_1 - \alpha_2)(\cos\theta \cdot \zeta_1^i + \sin\theta \cdot \zeta_2^i), \\ P_z^i &= E(\alpha_1 - \alpha_2)(\cos\theta \cdot \zeta_3^i + \sin\theta \cdot \zeta_4^i). \end{aligned} \right\} \tag{55}$$

Here, $\zeta_1^i = \sin^2\theta^i \cdot \sin^2\varphi^i - 1/3$, $\zeta_2^i = \sin^2\theta^i \sin\varphi^i \cdot \cos\varphi^i$, $\zeta_3^i = \sin\theta^i \cdot \cos\theta^i \cdot \sin\varphi^i$, and $\zeta_4^i = \sin\theta^i \cdot \cos\theta^i \cdot \cos\varphi^i$. Similar expressions can be obtained for the components of the moment induced in the j-th molecule. The quantities ζ_1^i, ζ_2^i, ζ_3^i, and ζ_4^i are functions of time because the angles ϑ and φ depend on time. Since the orientational motion of the i-th and j-th molecules is correlated, i.e., $\overline{\zeta_1^i(t')\zeta_2^j(t'')} \neq 0$, we can find the spectral density of the scattered-light intensity:

$$I_y(\Omega) = \frac{E^2}{\pi}(\alpha_1 - \alpha_2)^2 \int_0^\infty e^{i\Omega t} N \{\cos^2\theta \cdot \overline{\zeta_1(t')\,\zeta_1(t'')} + \sin^2\theta \cdot \overline{\zeta_2(t')\,\zeta_2(t'')}\}\,dt +$$

$$+ \frac{E^2}{\pi}(\alpha_1 - \alpha_2)^2 \int_0^\infty e^{i\Omega t} \sum_{\substack{i,\,j \\ i \neq j}} \cos\frac{2\pi d_{ij}}{\lambda} \{\cos^2\theta \cdot \overline{\zeta_1^i(t')\,\zeta_1^j(t'')} +$$

$$+ \sin^2\theta \cdot \overline{\zeta_2^i(t')\,\zeta_2^j(t'')} + 2\sin\theta \cdot \cos\theta \cdot \overline{\zeta_1^i(t')\,\zeta_2^j(t'')}\}\,dt. \qquad (56)$$

Here, d_{ij} is the distance between the i-th and j-th molecules; λ is the wavelength of the incident light; N is the number of the scattering molecules. It follows from this expression that the intensity of the forward-scattered light (angle θ) is higher than the intensity of the back-scattered light (angle $\pi - \theta$) because the last term becomes negative for $\pi - \theta$ provided $\overline{\zeta_1^i(t')\zeta_2^j(t'')} \neq 0$.

Unfortunately, the time dependences $\zeta_1(t)$ and $\zeta_2(t)$ are not known and therefore, it is not possible to integrate Eq. (56). Therefore, we shall consider only the integrated intensity. We shall replace ζ_1^i, ζ_2^i, ζ_1^i, and ζ_2^i, which represent the position of a molecule relative to the laboratory system of coordinates (x, y, z), with the angular functions ζ^i and ζ^j which represent the relative orientation of two molecules, by means of the formula

$$(\sigma'i) = \sum_\sigma (\sigma'\sigma)(i\sigma).$$

Here, σ and σ' are the axes of the coordinates linked to the i-th and j-th molecules, respectively; i = x, y, z. Next, averaging over all the orientations of the molecules, we obtain an expression for the integrated intensity of the light scattered by the anisotropy fluctuations:

$$\left.\begin{array}{l} I_y \propto E^2 \left\{\dfrac{2}{15}\cos^2\theta \left[\overline{(\zeta^i\zeta^j)^2} - \dfrac{1}{3}\right] + \dfrac{1}{10}\left[\overline{(\zeta^i\zeta^j)^2} - \dfrac{1}{3}\right] + \dfrac{1}{45}(3 + \cos^2\theta)\right\}, \\[2mm] I_z \propto E^2 \left\{\dfrac{1}{15} + \dfrac{1}{10}\left[\overline{(\zeta^i\zeta^j)^2} - \dfrac{1}{3}\right]\right\}. \end{array}\right\} \qquad (57)$$

The total intensity of the light scattered at an angle θ with respect to the incident wave is

$$I = I_y + I_z \propto \frac{E^2}{45}(6 + \cos^2\theta)\left(1 + \frac{3}{2}A\right), \qquad (58)$$

where

$$A = \overline{(\zeta^i\zeta^j)^2} - 1/3. \qquad (59)$$

If the orientations of the molecules are distributed at random, all the mean squares are given by $\overline{(\zeta^i\zeta^j)^2} = \frac{1}{3}$ and Eq. (58) reduces to Eq. (52) because in this case $\theta = 90°$ (θ is the angle between the electric field vector of the incident wave and the direction of observation of the scattered light).

Thus, the angular distribution of the intensity of the scattered light depends not only on the correlation in the orientational motion of the neighboring molecules but also on the existence of short-range orientational order.

The experimental results can be explained on the assumption that those liquids for which the ratio $z_1(\Omega)$ differs from unity exhibit a short-range orientational order and correlation in the rotational motion of neighboring molecules during short intervals of time $t < \tau_p$, i.e., intervals during which molecules are subject to rotational rocking.* If this assumption is correct, the light scattered by different molecules in a liquid is coherent and, therefore, the interference ensures that the forward scattering will predominate only in the high-frequency part of the Rayleigh wing.

In the case of carbon disulfide and carbon tetrachloride there is no such short-lived orientational order and, therefore, at all frequencies of the scattered light we find that $z_1(\Omega) = 1$.

The absence of asymmetry in the angular distribution of the intensity of the light scattered by viscous liquids at room temperature may mean, from the point of view of the theory of viscous liquids discussed here [66], that in the region where the correlation in the rotational motion of molecules is important the average orientation of the molecules is random although a short-range orientational order may appear in some parts of the region in question. When the temperature of a viscous liquid is raised, the number of correlated regions increases and a short-range orientational order may appear in these regions. This is probably why we find that in the case of salol at $t = 120°C$ the inequality $z_1(\Omega) > 1$ is obeyed.

The disagreement between our results and the angular dependence of the integrated intensity obtained by Cohen and Eisenberg [59] can be explained if we bear in mind that at frequencies corresponding to the predominant forward scattering the intensity is low. Therefore, such scattering makes a small contribution to the total intensity of the Rayleigh process and it could have been missed in [59]. A study of the angular distribution of the intensity in the wing of the Rayleigh line may provide a satisfactory method for investigating short-lived ($\sim 10^{-12}$-10^{-13} sec) orientational complexes of molecules in liquids. No other optical method is suitable for such studies.

§3. Experimental Investigation of the Wings of Depolarized Raman Lines of Liquids

We pointed out in §2 that the profiles of the depolarized Raman lines of liquids can be investigated experimentally up to 20 cm^{-1} from the line maximum. The use of a high-speed spectroscopic unit enabled us to investigate the profiles of six Raman lines of four liquids up to 80-90 cm^{-1}. Since it was of interest to determine to what extent the intensity distribution in the spectra of the depolarized Raman lines agreed with the intensity distribution in the wing of the Rayleigh line, the same apparatus was used also to study the Rayleigh wing of the four liquids in question.

The profiles of the depolarized lines and the wing of the Rayleigh line were investigated in benzene, carbon tetrachloride, chloroform, and toluene. These liquids were selected deliberately. In order to study the wings of the depolarized lines in the widest possible range of frequencies it was necessary to ensure that the lines in question were sufficiently strong in the Raman spectrum and that there were no other strong and broad lines within 100 cm^{-1} of those being investigated.

In the liquids selected the most suitable were the following Raman lines: 606 and 1178 cm^{-1} in the case of benzene, 313 cm^{-1} in the case of carbon tetrachloride, 261 and 761 cm^{-1} in the case of chloroform, and 217 cm^{-1} in the case of toluene. The degree of depolarization of these lines was close to $\frac{6}{7}$.

---

* This assumption is justified because $z_1(\Omega) > 1$ only in the far part of the wing, i.e., the part which is due to the rotational rocking of molecules.

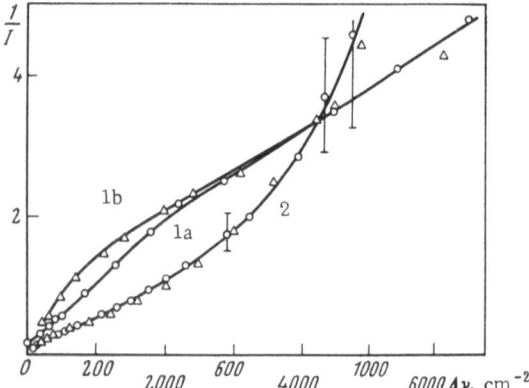

Fig. 6. Dependences, on the square of the frequency, of the reciprocals of the intensities in the wing of the Rayleigh line and in the profile of the $\nu = 261$ cm^{-1} Raman line of chloroform: 1a) near part of the Raman line (upper frequency scale); 1b) near part of the wing of the Rayleigh line; 2) far part of the wing of the Rayleigh line (triangles) and far part of the Raman line (circles, lower frequency scale).

Fig. 7. Dependences, on the square of the frequency, of the reciprocals of the intensities in the wing of the Rayleigh line and in the profile of the $\nu = 313$ cm^{-1} Raman line of carbon tetrachloride: 1) far part of the wing of the Rayleigh line (triangles); 2) far part of the Raman line (circles).

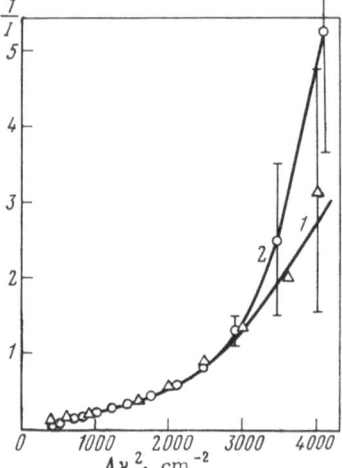

The distributions of the intensity in the wing of the Rayleigh line and in the $\nu = 261$ cm^{-1} Raman line of chloroform are plotted in Fig. 6 as the dependences $1/I = f(\Delta\nu^2)$. In the 25-80 cm^{-1} range the intensity distributions of both lines are identical (within the limits of the experimental error). In the near region (up to 25 cm^{-1}) the intensity in the wing of the $\nu = 261$ cm^{-1} line decreases more slowly than in the wing of the Rayleigh line. The profile of the Raman line, like the wing of the Rayleigh line, exhibits an inflection in the vicinity of $\Delta\nu \sim 15$ cm^{-1}, but the inflection is less pronounced in the Raman line. The same is true also of the second depolarized line of chloroform at $\nu = 761$ cm^{-1}.

A comparison of the distribution of intensity in the $\nu = 313$ cm^{-1} line profile with the intensity distribution in the wing of the Rayleigh line of the same liquid shows (Fig. 7) that the two profiles coincide. The discrepancy far from the line maximum is within the limits of the experimental error.

The Raman lines considered so far are due to nonplanar vibrations of molecules whose atoms are located in different planes. Somewhat different results are obtained when the wings of the Rayleigh lines of toluene and benzene are superimposed on the depolarized Raman lines of these liquids. A study of the $\nu = 217$ cm^{-1} line of toluene, corresponding to nonplanar vibrations, shows that the coincidence described above is no longer obtained (Fig. 8). The intensity of the wing of the $\nu = 217$ cm^{-1} line decreases (far from the maximum) much faster than the intensity of the wing of the Rayleigh line. On the other hand, in the near part of the spectrum the profile of the Raman line is less steep than that of the Rayleigh line.

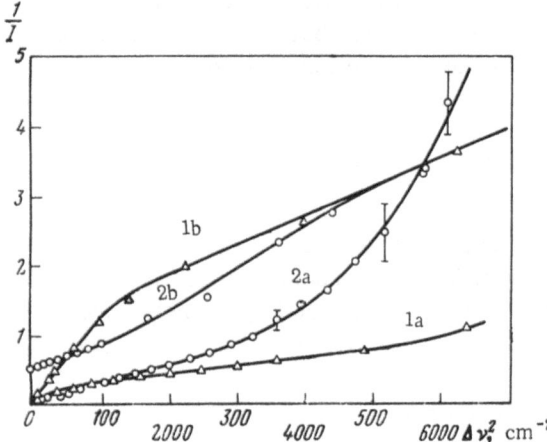

Fig. 8. Dependences, on the square of the frequency, of the reciprocals of the intensities in the wing of the Rayleigh line and in the profile of the $\nu = 217$ cm^{-1} Raman line of toluene: 1a) far part of the wing of the Rayleigh line (triangles); 2a) far part of the Raman line (circles, lower frequency scale); 1b) near part of the wing of the Rayleigh line (triangles); 2b) near part of the Raman line (circles, upper frequency scale).

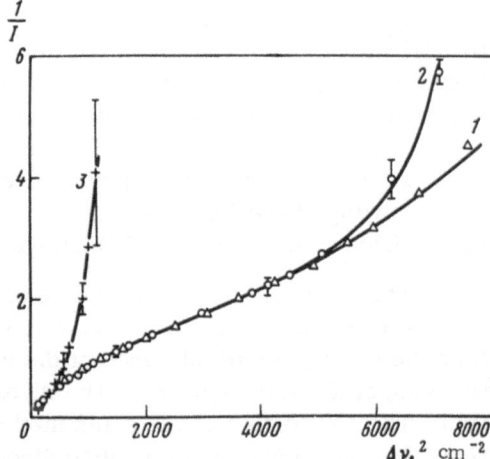

Fig. 9. Dependences, on the square of the frequency, of the reciprocals of the intensities in the wing of the Rayleigh line and in the profiles of the $\nu = 1178$ and 606 cm^{-1} Raman lines of benzene: 1) far part of the wing of the Rayleigh line (triangles); 2) far part of the $\nu = 1178$ cm^{-1} Raman line (circles); 3) far part of the $\nu = 606$ cm^{-1} Raman line (crosses).

We shall now compare the wings of the $\nu = 606$ and 1178 cm^{-1} Raman lines of benzene with the wing of the Rayleigh line. The two Raman lines correspond to planar vibrations of the molecules. In a fairly wide range of frequencies the profile of the 1178 cm^{-1} line coincides with the wing of the Rayleigh line; only in the far region, beginning from 75 cm^{-1} measured from the line maximum, there is some discrepancy which is larger than the experimental error (Fig. 9). This is not true of the other line of benzene at $\nu = 606$ cm^{-1}. The intensity of its wing decreases so rapidly, compared with the intensities of the Rayleigh line and of the depolarized $\nu = 1178$ cm^{-1} line, that a considerable discrepancy is observed at frequencies of 25 cm^{-1}.

A characteristic feature of all the investigated depolarized Raman lines is a fairly large width in the near (up to 25 cm^{-1}) part of the spectrum, compared with the width of the wing of the Rayleigh line of the same liquids. Moreover, in the case of liquids whose molecules have different moments of inertia (benzene, toluene), the intensity in the high-frequency part of the Raman scattering spectrum decreases much faster than in the corresponding part of the Rayleigh wing.

The first of these features can be explained as follows. The near region (up to 10-20 cm^{-1}) of the wing of the Rayleigh line of low-viscosity liquids is due to orientational jumps of molecules from one equilibrium position to another. The width of this part of the wing is determined simply by the rate of these molecular jumps. The width of the depolarized Raman lines is influenced not only by the Brownian rotation of the molecules but also by the dissipative loss of vibrational quanta from the molecules and by the interaction between the given molecule and its nearest neighbors. It is the influence of these additional broadening mechanisms that is responsible for the slower fall of the intensity in the profiles of the Raman lines, compared with the corresponding fall in the intensity in the wing of the Rayleigh line.

The high-frequency part of the wing of the Rayleigh line is due to the modulation of the scattered light by the rotational rocking of molecules about their temporary equilibrium positions. The wing of the Rayleigh line includes contributions from the rocking of molecules about their principal axes which may have different moments of inertia. A good agreement throughout the high-frequency part of the wing of the Rayleigh line and the depolarized $\nu = 261$ and 761 cm^{-1} Raman lines of chloroform, the $\nu = 313$ cm^{-1} line of carbon tetrachloride, and $\nu = 1178$ cm^{-1} line of benzene suggests a common broadening mechanism.

The noncoincidence of the profiles of the wing of the Rayleigh line and of the Raman line in the case of nonplanar vibrations of toluene ($\nu = 217$ cm^{-1}) can be explained by the easier rocking of this molecule about an axis perpendicular to its plane than about the axes passing through that plane. The far part of the wing of the Rayleigh line is influenced by all the rotational rocking motions of the molecules but an important contribution is made by the rocking around the axis which is perpendicular to the plane of the molecule. However, such rocking cannot modulate the $\nu = 217$ cm^{-1} Raman line because it corresponds to nonplanar vibrations. This is probably why the far part of the profile of the $\nu = 217$ cm^{-1} line shows a steeper fall than the wing of the Rayleigh line. It is not yet possible to account for the noncoincidence of the profiles of the wing of the Rayleigh line and of the $\nu = 606$ cm^{-1} Raman line of benzene.

Thus, a combination of studies of the wing of the Rayleigh line and of profiles of the depolarized Raman lines — carried out in a wide range of frequencies — makes it possible to determine the degree of influence of the rotational rocking of molecules about various axes on the wing of the Rayleigh line. The investigation reported above is the first step in this direction. The application of the same method to solutions looks very promising because in solutions the molecules can be studied directly.

CHAPTER IV

RESULTS OF AN INVESTIGATION OF THE STIMULATED MOLECULAR SCATTERING OF LIGHT

§1. Influence of Temperature on the Stimulated Scattering in the Wing of the Rayleigh Line

The stimulated scattering in the wing of the Rayleigh line was observed in benzaldehyde, ortho-xylene, quinoline, ortho-nitrotoluene, and acetophenone. All these liquids consist of strongly anisotropic molecules.

The spectrum of the stimulated wing of the Rayleigh line of ortho-xylene, recorded at room temperature, is shown in Fig. 10b. The emission spectrum of a ruby laser is given for comparison in Fig. 10a. When giant laser pulses of ~ 100 MW power were used, it was found that the stimulated Stokes wing of ortho-xylene occupied a region of 2.5 cm^{-1} width and the distribution of intensity in this wing had a maximum at ~ 0.7 cm^{-1} [67]. However, the position of this maximum was difficult to determine accurately because the wing was fairly wide, and,

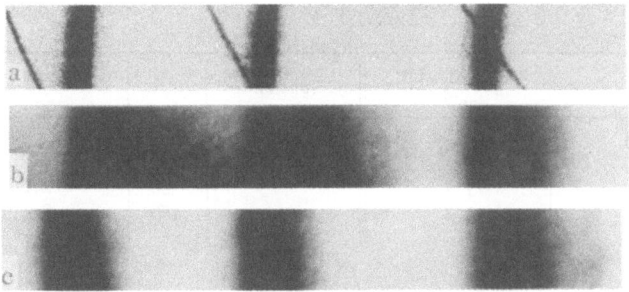

Fig. 10. Spectra of the exciting radiation (a) and of
the stimulated wing of the Rayleigh line of ortho-
xylene (b) and quinoline (c). Spectral range of the
interferometer 5 cm^{-1}.

moreover, it overlapped two components of the stimulated Brillouin scattering. The position
of the incipient maximum in the stimulated wing of ortho-xylene was in agreement with the experimental data [41] on the half-width of the thermal wing of the Rayleigh line ($\Delta\nu \sim 0.65$ cm^{-1}).

In the case of quinoline (Fig. 12c) the stimulated wing recorded at 80°C was in the form
of a strong well-resolved band in the Stokes region, extending to ~ 1.7 cm^{-1} from the exciting
line. The Brillouin components were absent from the scattered-light spectrum and no maximum was found in the stimulated wing.

In the case of ortho-nitrotoluene, acetophenone, and other liquids, we observed only the
Stokes component of the stimulated wing. At 20°C this component occupied ~ 3 cm^{-1}. The wing
was overlapped by six Brillouin components and therefore the expected maximum was difficult
to detect.

A study of the influence of temperature on the stimulated scattering in the wing of the
Rayleigh line was carried out for nitrobenzene (Fig. 11) and benzaldehyde (Fig. 12). It was
found that the width of the stimulated wing increased smoothly with increasing temperature.
The results of measurements are presented in Table 2.

This temperature dependence of the width of the stimulated wing was expected because
the anisotropy relaxation time decreased with increasing temperature [Eq. (3)] and, consequently, the gain maximum shifted in the Stokes direction away from the undisplaced line
[Eq. (35)].

The stimulated wing of nitrobenzene had a broad maximum at ~ 0.5 cm^{-1} from the undisplaced line. As in the case of ortho-xylene, the position of this maximum corresponded to the
half-width of the thermal wing of the Rayleigh line ($\Delta\nu = 0.45$ cm^{-1}). In the case of benzalde-

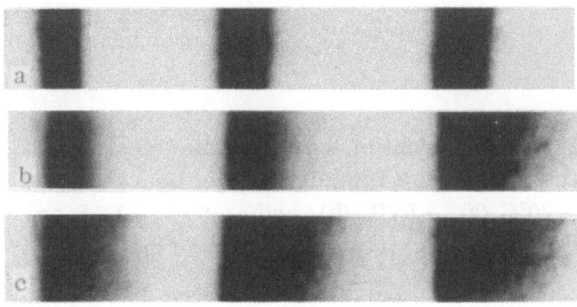

Fig. 11. Spectrum of the stimulated scattering
in the wing of the Rayleigh line of nitrobenzene
at -20°C (a), -65°C (b), and -120°C (c). Spectral range of the interferometer 5 cm^{-1}.

TABLE 2

Substance	t, °C	Width of wing, cm^{-1}	Substance	t, °C	Width of wing, cm^{-1}
Benzaldehyde	20	1	Nitrobenzene	20	1.2
	57	1.7		65	2.2
	100	2.2		120	3

hyde the maximum was not observed: it was most probably masked by the everpresent Brillouin components. However, the absence of a definite maximum in the case of quinoline, nitrotoluene, acetophenone, and benzaldehyde could also be due to the fact that, when the gain g_1 was sufficiently large for the observation of the stimulated wing, the field was so high that the condition of validity of the relevant theory was no longer obeyed [8, 52]:

$$\frac{\alpha_1 - \alpha_2}{kT} E^2 \ll 1.$$

Heating of benzaldehyde reduced the number of the Brillouin components superimposed on the stimulated wing. In all liquids a gradual reduction of the laser power resulted in a rapid fall of the intensity of the stimulated wing and its eventual disappearance. In some cases (ortho-xylene) a reduction in the intensity of the laser radiation by a pile of glass plates revealed several Brillouin components in place of the Stokes wing and, consequently, the stimulated Brillouin scattering was quenched (at least in some liquids) by the appearance of the stimulated wing of the Rayleigh line, in spite of the fact that the threshold for the appearance of the Brillouin scattering was lower than the threshold of the stimulated wing. Obviously the Brillouin scattering and the stimulated wing processes were in competition.

We pointed out in §2 of Chap. II that special measures were taken to avoid amplification of the Brillouin components in the ruby laser. A quarter plate P_1, a polarizer, or a glass plate inclined at the Brewster angle P_0 (Fig. 2) weakened regenerative amplification of the back-scattered light. In spite of these measures we found that several (usually two) Brillouin components remained in the scattered-light spectra of most of the liquids. Since only the forward-scattered light could be observed in our apparatus (the back-scattered light was solely due to reflections), we concluded that the first Brillouin component appeared because of the reflection from the front window of the cuvette, from the end of the ruby crystal, etc. The second Brillouin component was primarily due to the local multiple scattering in the nonlinear interaction of light waves with hypersound. In this local scattering process a high-power light wave, which is back-scattered and displaced by Ω_B in its frequency, generates a new light wave inside the interaction region. The new wave is shifted by $2\Omega_B$ and travels in the original direction of the exciting light, as demanded by the law of conservation of momentum of a photon scattered by a hyper-

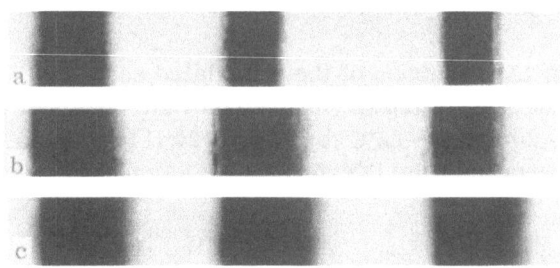

Fig. 12. Spectrum of the stimulated scattering in the wing of the Rayleigh line of benzaldehyde at -20°C (a), -57°C (b), and -100°C (c). Spectral range of the interferometer 5 cm^{-1}.

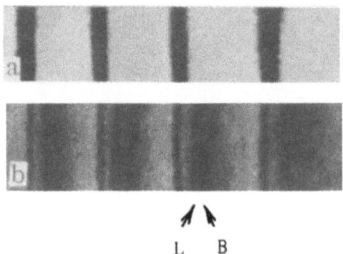

Fig. 13. Spectra of the exciting line (a) and of the stimulated Brillouin scattering in nitrobenzene (b). B is the Brillouin component; L is the amplified mode of the laser radiation. Spectral range of the interferometer 1 cm^{-1}.

sonic wave. This mechanism is supported by the results of our experiments in which the scattering region was separated by 30 m from the laser. Since the duration of a giant pulse did not exceed 15 nsec and a second pulse was not produced for 100 nsec, the light scattered back to the laser could not have been amplified in the ruby crystal. In spite of this we always observed two Brillouin components.

The local multiple scattering in the nonlinear interaction between light and hypersonic waves can also explain the appearance of the anti-Stokes components in the Brillouin region [18].

The spectrograms of the scattered light obtained in the earlier investigations [16, 68] and in the present study included a narrow and fairly strong line,* which could be attributed neither to the stimulated Brillouin scattering nor to the stimulated wing (Fig. 13). The line was absent from the spectrum of the exciting radiation. Its separation from the undisplaced line was 0.09-0.13 cm^{-1} at room temperature and it increased weakly when the temperature of the liquid was raised. For example, in the case of quinoline at 5°C this line was separated by 0.11 cm^{-1} from the undisplaced line whereas at 95°C this separation was 0.12 cm^{-1}. The intensity of this new line was so high that it sometimes produced its own Brillouin component. When the exciting radiation was attenuated to a level at which the wide part of the stimulated wing disappeared, this narrow line still remained. We established that the line was a weak laser mode because the separation between this line and the principal mode changed by a factor of 1.5 when the distance between the resonator mirrors was altered by the same factor. When this mode reached a medium with anisotropic molecules, it was amplified by practically the same mechanism which was responsible for the stimulated wing.

A similar amplification of a laser mode in nitrobenzene was recently observed by Cho, Foltz, Rank, and Wiggins [55], who took this line to be a peak in the stimulated wing of the Rayleigh line. In their experiments the second laser mode was so weak that it did not appear in the spectrum of the exciting radiation. However, it could be amplified during its propagation in a medium containing anisotropic molecules. The observations of Cho et al. can also be explained by a mechanism similar to that suggested by Brewer [69] for the stimulated Brillouin scattering. The light corresponding to the stimulated wing of nitrobenzene is back-scattered to the ruby laser and it excites there a mode with the highest Q factor. When the liquid is heated, the peak value of the gain of the wing shifts in the Stokes direction and this gives rise to a temperature dependence of the position of the observed mode.

It is interesting to note that self-focusing is observed in all the liquids which exhibit the stimulated wing [70]. The correlation between the two effects is quite natural. In fact, the stimulated wing of the Rayleigh line appears because of orientation of anisotropic molecules in a liquid by the total electric field of the light wave. Such molecular orientation has a non-linear effect on the refractive index and, therefore, light beams collected at the focus of a lens are pulled into a channel from which they cannot escape because of total internal reflection at the channel's boundary. This self-focusing process increases considerably the energy density

---

* This line was located between the exciting line and the Brillouin component, as shown in Fig. 1 obtained by Mash et al. [68].

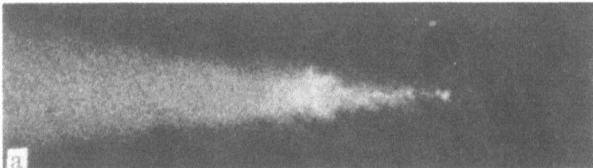

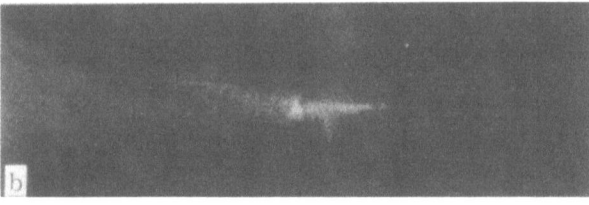

Fig. 14. Self-focusing of a laser beam in quino-line: a) initial intensity of the giant pulse; b) intensity reduced by a factor of 1.5; c) intensity reduced by a factor of 5. The direction of the beam is from the left to the right.

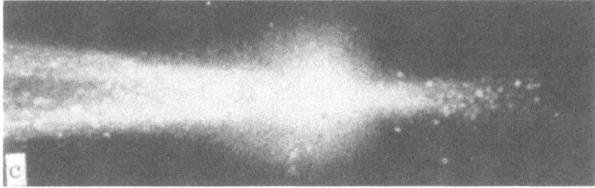

in the channel and, consequently, it enhances the stimulated scattering of the wing of the Rayleigh line.

In some liquids (for example, quinoline) the laser beam disappeared beyond the focus if a short-focus lens (f = 3 cm) was used. The energy density at the focus was so high that a spark appeared at that point and much of the laser beam energy was absorbed by the spark-generated plasma (Fig. 14a). When the incident radiation power was reduced, the spark at the focus disappeared and a clearly visible channel was observed beyond the focus (Fig. 14b). When further attenuation was applied to the laser beam, the propagation of light was found to be in accordance with the laws of geometrical objects.

The considerable increase in the energy density resulting from self-focusing could be responsible for the great changes in the nature of the stimulated Raman scattering reported in [71, 72]. Lallemand and Bloembergen [71] showed by direct experiments that the critical length needed to produce the stimulated Raman scattering was reduced strongly by insertion of an additional cuvette containing a strongly self-focusing liquid (bromobenzene) between the laser and a cuvette containing nitrobenzene.

Self-focusing can affect not only the threshold but also the profile of the stimulated Raman scattering. In some cases [72-74] an anomalous broadening of the Rayleigh and the Raman scattering lines is observed under self-focusing conditions.

Aref'ev, Fabelinskii, Kyzylasov, Starunov, and the present author [75] observed the Stokes and anti-Stokes components of the ν = 658 cm^{-1} stimulated Raman line as well as the first overtones of these components when they studied carbon disulfide in which self-focusing occurred extremely easily. All the Raman lines as well as the Rayleigh line had stimulated wings extending up to 100 cm^{-1} in the Stokes region. The appearance of such wide and strong wings near the Raman lines could be due to two causes. The total light field of the Raman line and its thermal wing could be enhanced by self-focusing so much that it could give rise to a stimulated wing, as observed in the Rayleigh scattering process. Alternatively, if the intensity of the wing of the Rayleigh line exceeded the threshold for the stimulated Raman scattering, the wing near the 658 cm^{-1} line was simply a continuous spectrum of Raman lines excited by the components of the Rayleigh wing. In this case the self-focusing would play the decisive role.

§2. Four-Photon Interaction in the Stimulated Wing of the Rayleigh Line

The four-photon interaction was first discovered in the stimulated Raman scattering by Maker and Terhune [76].

We shall now describe the results of observation of a four-photon interaction in the stimulated wing of the Rayleigh line (low-angle scattering), i.e., the scattering of light on light in an optically nonlinear medium. The observed effect is a special case of parametric interaction between light waves in a nonlinear medium consisting of anisotropic molecules, which is observed when the angle between the directions of propagation of these waves is small [52]. The theory of this effect is considered in §3 of Chap. I.

We recall that in the four-photon interaction in the stimulated wing of the Rayleigh line the maximum gain of the Stokes and anti-Stokes components of the wing corresponding to the optimal angle $\vartheta_{opt} = \pm |E_0| A^{1/2}$ is given by

$$g_2(\Omega) = -2k_\omega + A |k_1| |E_0|^2 (1 + \Omega^2 \tau^2)^{-1/2},$$

where k_ω is the amplitude coefficient of absorption of light in the medium in question; k_1 is the wave vector of the scattered light; Ω is the frequency measured from the exciting lines; τ is the anisotropy relaxation time. It is evident from the above expression that the gain maximum occurs at $\Omega = 0$. The existence of the four-photon interaction in the stimulated wing of the Rayleigh line should be manifested by a wing which appears when the angle is optimal ϑ_{opt} (this angle is small because of smallness of the ratio $A = \varepsilon_2/\varepsilon_0$). This wing should be the same on the Stokes and anti-Stokes sides and it should have its intensity peak at the frequency of the exciting line. If there is no interaction between the Stokes and anti-Stokes light waves, only the Stokes wing is observed and the gain maximum occurs at $\Omega = 1/\tau$ (the anti-Stokes wave is not amplified).

We investigated the four-photon interaction using a ruby laser emitting pulses of ~ 90 MW power. The laser radiation was focused inside a liquid-filled cuvette by a short-focus lens ($f = 2.5$ cm). The stimulated wing of the Rayleigh line was observed at angles of $\theta \sim 0$ and $90°$.

As expected the anti-Stokes wing of ortho-xylene and nitrobenzene was observed only in some of the interferometric orders corresponding to the optimal angles. In our experiments these angles did not exceed $2°$. The spectrum of the stimulated wing of the Rayleigh line of nitrobenzene is shown in Fig. 15b. The Stokes and anti-Stokes parts of the wing can be seen on the right. The part of the interferogram on the right corresponds to the space within the

Fig. 15. Spectra of the exciting light (a) and the stimulated wing of the Rayleigh line (b) under four-photon interaction conditions in nitrobenzene. S is the Stokes wing; AS is the anti-Stokes wing. Spectral range of the interferometer 8.33 cm^{-1}.

cone formed by the optimal angle. The part on the left corresponds to the region outside this cone and, therefore, the anti-Stokes wing is absent on the left.

When the scattering angle was $\theta = 90°$, only the Stokes part of the wing was observed in the spectrum of the stimulated scattering of light. A rough estimate of the gain corresponding to the stimulated wing gave $g_2 \sim 10^3$. The region occupied by the anti-Stokes component was ~ 1.5 cm^{-1}. Under the four-photon interaction conditions the Stokes component of the wing was somewhat wider (~ 2 cm^{-1}). When the scattering angle was $\theta > \vartheta_{opt}$, the Stokes component occupied only ~ 1 cm^{-1} (the left-hand side of Fig. 15b). This difference between the widths of the Stokes component in the presence and absence of the four-photon interaction was expected because the gain of the Stokes wing $g_2(\Omega)$ was greater than $g_1(\Omega)$. As expected, the maximum in the stimulated wing observed under the four-photon interaction conditions was located at a frequency equal to the frequency of the exciting light.

In the preceding section we mentioned that the stimulated scattering of light in the wing of the Rayleigh line could suppress the stimulated Brillouin scattering. Such suppression was even stronger when the anti-Stokes wing was present.

A special case of the four-photon interaction in nitrobenzene corresponding to $\Omega = 0$ and $g \sim 10$ was observed by Carman, Chiao, and Kelley [77]. They focused giant laser pulses in a nitrobenzene-filled cuvette and directed a second weak wave at an angle of $+\vartheta_{opt}$ with respect to the direction of the main beam (the weak wave was simply part of the main beam split by a system of plates and mirrors). When the four-photon interaction took place, additional radiation was observed at an angle of $-\vartheta_{opt}$.

If the four-photon interaction occurs in the stimulated wing of the Rayleigh line, the anti-Stokes components of the Brillouin scattering may appear as a result of this interaction and of the local multiple scattering in the interaction region (mentioned in the preceding section).

Thus, the four-photon interaction in the stimulated scattering of light corresponding to the wing of the Rayleigh line can provide an effective method for the determination of nonlinear properties of various media.

The author is deeply grateful to his scientific supervisors I. L. Fabelinskii and V. S. Starunov for suggesting the subject and for their constant interest during the investigation. He is also grateful to I. I. Sobel'man for discussing the results obtained and to I. Ya. Dikhter, V. P. Zaitsev, Yu. I. Kyzylasov, and V. V. Morozov for their help in the experiments.

LITERATURE CITED

1. I. L. Fabelinskii, Molecular Scattering of Light, Plenum Press, New York (1968).
2. M. V. Vol'kenshtein, Molecular Optics [in Russian], Gostekhizdat, Moscow (1951).
3. M. A. Leontovich, J. Phys. USSR, 4:499 (1941).
4. S. M. Rytov, Zh. Eksp. Teor. Fiz., 33:514, 671 (1957).
5. J. Cabannes and P. Daure, C. R. Acad. Sci., 186:1533 (1928).
6. C. V. Raman and K. S. Krishnan, Nature, 122:278, 882 (1928).
7. V. S. Starunov, E. V. Tiganov, and I. L. Fabelinskii, ZhETF Pis. Red., 5:317 (1967).
8. V. S. Starunov, Tr. Fiz. Inst. Akad. Nauk SSSR, 39:151 (1967).
9. J. Frenkel, Kinetic Theory of Liquids, Oxford University Press (1946).
10. I. L. Fabelinskii, Tr. Fiz. Inst. Akad. Nauk SSSR, 9:181 (1958).
11. V. S. Starunov, Dokl. Akad. Nauk SSSR, 153:1055 (1963).
12. V. S. Starunov, Opt. Spektrosk., 18:300 (1965).
13. R. Y. Chiao, C. H. Townes, and B. P. Stoicheff, Phys. Rev. Lett., 12:592 (1964).

14. E. Garmire and C. H. Townes, Appl. Phys. Lett., 5:84 (1964).

15. R. G. Brewer and K. E. Rieckhoff, Phys. Rev. Lett., 13:334a (1964).

16. D. I. Mash, V. V. Morozov, V. S. Starunov, and I. L. Fabelinskii, ZhETF Pis. Red., 2:41 (1965).

17. G. I. Zaitsev, Yu. I. Kyzylasov, V. S. Starunov, and I. L. Fabelinskii, ZhETF Pis. Red., 6:802 (1967).

18. R. V. Wick, D. H. Rank, and T. A. Wiggins, Phys. Rev. Lett., 17:466 (1966).

19. G. I. Zaitsev, Opt. Spektrosk., 23:325 (1967).

20. G. I. Zaitsev and V. S. Starunov, Opt. Spektrosk., 19:893 (1965).

21. G. I. Zaitsev and V. S. Starunov, ZhETF Pis. Red., 4:54 (1966).

22. G. I. Zaitsev, Yu. I. Kyzylasov, V. S. Starunov, and I. L. Fabelinskii, ZhETF Pis. Red., 6:505 (1967).

23. G. I. Zaitsev, Yu. I. Kyzylasov, V. S. Starunov, and I. L. Fabelinskii, ZhETF Pis. Red., 6:695 (1967).

24. I. L. Fabelinskii, Izv. Akad. Nauk SSSR, Ser. Fiz., 11:382 (1947).

25. J. Cabannes and Y. Rocard, J. Phys. Radium, 10:52 (1929).

26. V. L. Ginzburg, Dokl. Akad. Nauk SSSR, 30:399 (1941).

27. E. Gross, Nature, 126:201, 400, 603 (1930).

28. C. S. Venkateswaran, Proc. Indian Acad. Sci., A8:448 (1938).

29. L. D. Landau and G. Placzek, Phys. Z. Sowjetunion, 5:172 (1934).

30. P. Debye, Polar Molecules, Chemical Catalog Co., New York (1929) [reprinted by Dover, New York (1947)].

31. I. L. Fabelinskii, Opt. Spektrosk., 2:510 (1957).

32. Yu. N. Zhivlyuk, Dissertation [in Russian], Moscow Physicotechnical Institute and Physics Institute, Academy of Sciences of the USSR, Moscow (1960).

33. V. L. Ginzburg, Zh. Eksp. Teor. Fiz., 34:246 (1958).

34. E. F. Gross, Izv. Akad. Sci. SSSR, Ser. Fiz., 5:19 (1941).

35. H. A. Kramers, Physica, 7:284 (1940).

36. R. A. Sack, Proc. Phys. Soc. London, B70:414 (1957).

37. M. F. Vuks and V. L. Litvinov, Dokl. Akad. Nauk SSSR, 105:696 (1955).

38. M. F. Vuks, Opt. Spektrosk., 9:92 (1960).

39. A. K. Atakhodzhaev, Izv. Akad. Nauk Uzbek SSR, Ser. Fiz.-Mat. Nauk, No. 1, p. 86 (1962).

40. A. I. Chernyavskaya and G. P. Roshchina, Vestn. Leningrad. Univ., Fiz. Khim., No. 16, p. 26 (1964).

41. A. K. Atakhodzhaev, M. F. Vuks, and V. L. Litvinov, Proc. Tenth Conf. on Spectroscopy [in Russian], Vol. 1, L'vov (1967).

42. G. I. Zaitsev and V. S. Starunov, Opt. Spektrosk., 22:409 (1967).

43. I. I. Sobel'man, Izv. Akad. Nauk SSSR, Ser. Fiz., 17:554 (1953).

44. K. A. Valiev, Zh. Eksp. Teor. Fiz., 40:1832 (1961).

45. K. A. Valiev, Opt. Spektrosk., 11:465 (1961).

46. K. A. Valiev and L. D. Éskin, Opt. Spektrosk., 12:758 (1962).

47. A. V. Rakov, Opt. Spektrosk., 13:369 (1962).

48. A. V. Rakov, Opt. Spektrosk., 7:202 (1959).

49. A. V. Rakov, Tr. Fiz. Inst. Akad. Nauk SSSR, 27:111 (1964).

50. A. I. Sokolovskaya, Tr. Fiz. Inst. Akad. Nauk SSSR, 27:63 (1964).

51. N. I. Rezaev, Proc. Tenth Conf. on Spectroscopy [in Russian], Vol. 1, L'vov (1967).

52. V. S. Starunov, Dokl. Akad. Nauk SSSR, 179:65 (1968).

53. R. Y. Chiao, P. L. Kelley, and F. Farmire, Phys. Rev. Lett., 17:1158 (1966).

54. N. Bloembergen and P. Lallemand, Phys. Rev. Lett., 16:81 (1966).

55. C. W. Cho, N. D. Foltz, D. H. Rank, and T. A. Wiggins, Phys. Rev. Lett., 18:107 (1967).

56. M. L. Sosinskii, Izv. Akad. Nauk SSSR, Ser. Fiz., 17:621 (1953).

57. S. Tolansky, High Resolution Spectroscopy, Methuen, London (1947).
58. T. S. Velichkina, Tr. Fiz. Inst. Akad. Nauk SSSR, 9:59 (1958).
59. G. Cohen and H. Eisenberg, J. Chem. Phys., 43:3881 (1965).
60. G. I. Pokrovskii (Pokrovski) and E. A. Gordon, Ann. Phys. (Leipzig), 4:488 (1930).
61. A. I. Sokolovskaya and P. D. Simova, Opt. Spektrosk., 15:622 (1963).
62. I. I. Kondilenko, P. A. Korotkov, and V. L. Strizhevskii, Opt. Spektrosk., 11:169 (1961).
63. L. D. Landau and E. M. Lifshitz, Electrodynamics of Continuous Media, Pergamon Press, Oxford (1960).
64. G. S. Landsberg, P. A. Bazhulin, and M. M. Sushchinskii, Principal Parameters of Raman Spectra of Hydrocarbons [in Russian], Izd. AN SSSR, Moscow (1956).
65. K. A. Stacey, Light-Scattering in Physical Chemistry, Academic Press, New York (1956).
66. M. A. Isakovich and I. A. Chaban, Dokl. Akad. Nauk SSSR, 165:299 (1965).
67. I. Ya. Dikhter, Dissertation [in Russian], Moscow Physicotechnical Institute and Physics Institute, Academy of Sciences of the USSR, Moscow (1967).
68. D. I. Mash, V. V. Morozov, V. S. Starunov, E. V. Tiganov, and I. L. Fabelinskii, ZhETF Pis. Red., 2:246 (1965).
69. R. G. Brewer, Appl. Phys. Lett., 9:51 (1966).
70. G. A. Askar'yan, Zh. Eksp. Teor. Fiz., 42:1567 (1962).
71. P. Lallemand and N. Bloembergen, Phys. Rev. Lett., 15:1010 (1965).
72. S. A. Akhmanov, A. I. Sukhorukov, and R. V. Khokhlov, Usp. Fiz. Nauk, 93:19 (1967).
73. T. K. G. Gustafson, F. DeMartini, C. H. Townes, and P. L. Kelley, Bull. Amer. Phys. Soc., 12:687 (1967).
74. P. Lallemand, Appl. Phys. Lett., 8:276 (1966).
75. I. M. Aref'iev, I. L. Fabelinskii, Yu. I. Kyzylasov, V. S. Starunov, and G. I. Zaitsev (Zaitzev), Phys. Lett., 26A:82 (1967).
76. P. D. Maker and R. W. Terhune, Phys. Rev., 137:A801 (1965).
77. R. L. Carman, R. Y. Chiao, and P. L. Kelley, Phys. Rev. Lett., 17:1281 (1966).

LIGHT SCATTERING STUDIES OF THE PROPAGATION OF LONGITUDINAL AND TRANSVERSE HYPERSONIC WAVES IN LIQUIDS *

E. V. Tiganov

The positions and widths of the Brillouin scattering components were used in a determination of the velocity and absorption of the longitudinal hypersonic wave of 10^9-10^{10} Hz frequency in liquids. The first observations were made of a fine structure in the wing of the Rayleigh line of liquids. The measured separations between the fine structure components were used to determine the velocity of the transverse hypersonic waves and of the shear moduli of nitrobenzene, quinoline, and aniline at 10^9 Hz. A helium–neon gas laser was used as the excitation source and a Fabry–Perot interferometer as the spectroscopic instrument.

INTRODUCTION

Molecular scattering of light is an important branch of physical optics. It is used widely in studies of the structure of molecules; in molecular physics, physical chemistry, physics, and chemistry of polymers, proteins, and electrolytes; in physical acoustics and other branches of science [1].

The present paper is concerned solely with the molecular scattering of light which is due to fluctuations of the refractive index (or the permittivity) resulting from fluctuations of the density and anisotropy.

Random fluctuations in a medium vary continuously in space and time, they appear and disappear, and different fluctuations vary differently with time.

Such time-varying fluctuations not only scatter light but also alter considerably its spectral composition. For example, different types of variation of fluctuations with time give rise to different spectral compositions of the light scattered by molecular processes. Therefore, investigations of the spectra of the molecular scattering of light in different media under different conditions can give much valuable and sometimes unique information on transport processes and particularly on the propagation of hypersound.

The first theoretical investigations of the molecular scattering of light by adiabatic fluctuations of the density were carried out by Mandel'shtam [2] and Brillouin [3]. It is worth

* Condensed text of a thesis submitted for the degree of Candidate of Physicomathematical Sciences, defended on February 26, 1968, at the P. N. Lebedev Physics Institute, Academy of Sciences of the USSR. Scientific supervisors: I. L. Fabelinskii and V. S. Starunov.

stressing that Mandel'shtam also considered the spectrum of the molecular scattering of light resulting from isobaric fluctuations of the density.

Temporal and spatial adiabatic fluctuations of the density or pressure can be described by the wave equation. Therefore, we may regard the scattered light as modulated by a periodic function and we can expect the appearance of discrete Brillouin components in the scattered light.*

The shift of the Brillouin components with respect to the exciting line can be used to determine the velocity of hypersound or of that Debye wave which — for a given scattering angle and wavelength of the exciting radiation (satisfying the Bragg condition) — is responsible for the appearance of the Brillouin doublet. The width of the doublet lines can be used to determine the amplitude absorption coefficient of hypersound.

The frequency of hypersound in the liquids investigated in the present study lies in the range 10^9-10^{10} Hz.

Time-varying isobaric fluctuations of the density or entropy can be described by the heat conduction equation. The modulation of scattered light by time-varying fluctuations of the entropy broadens the central component and the broadening increases with the thermal diffusivity. This central component of the fine structure of the Rayleigh line will not be considered in the present paper.

Scattered light is also modulated by the extremely rapid fluctuations of the anisotropy or orientation of the axes of anisotropy molecules. This process is extremely complex and represented by the wide background surrounding the Rayleigh line: it is known as the wing of the Rayleigh line.

The first rational explanation of the nature of the spectrum of light scattered by fluctuations of the anisotropy was given by Landau and Placzek [4]. They suggested that the wing of the Rayleigh line was due to the relaxation process responsible for the dispersion of electromagnetic waves in polar liquids and the half-width of the wing was governed by the reciprocal of the relaxation time of the dipole moment. In fact, the situation is much more complex.

Theoretical and experimental investigations of Leontovich [5], Fabelinskii [6], Rytov [7], and Starunov [8] established the principal relationships governing the distribution of intensity in the wing of the Rayleigh line and provided a theoretical explanation which applies to low-viscosity liquids.

The experimental studies of Fabelinskii [6] showed that the scattering of light in some liquids gives rise to a wing which is so narrow (in the region of highest intensity) that it can be observed in the interference spectrum. The half-width of this wing amounts to a few tenths of a reciprocal centimeter.

The present paper reports an experimental study of this narrow wing of the Rayleigh line. This study was carried out using an improved technique and new substances. A fine structure in the wing of the Rayleigh line was observed for the first time.

Up to five years ago all the investigations of the fine structure of the wing of the Rayleigh line were carried out using gas-discharge lamps usually filled with mercury. The narrow spectral lines necessary for the excitation of the scattered light were emitted by gas-discharge lamps at very low pressures and were separated from the rest of the spectrum by a monochromator. Consequently, these lines were very weak. However, in spite of these dif-

---

*Brillouin scattering is usually referred to as the Mandel'shtam–Brillouin scattering in the
  Soviet literature.

ficulties, the fine structure of the central Rayleigh line was first observed with the aid of mercury radiation [9-11]. The results obtained in these experiments were generally in satisfactory
agreement with the theoretical predictions relating to the splitting of this line [2, 3].

Mandel'shtam and Leontovich [12] developed a relaxation theory of the propagation of
sound in condensed media. This theory overcame several difficulties encountered in the classical treatments and it predicted the existence of a strong dispersion of the velocity of sound.

Fabelinskii and his colleagues [13-15] detected considerable dispersion in the velocity
of sound in low-viscosity liquids by comparing the measured velocity of hypersound with the
positions of the Brillouin components of the Rayleigh line. This dispersion is due to relaxation of the bulk viscosity.

Investigations of Fabelinskii and Pesin [16, 17] of liquids with a high shear viscosity revealed a fine structure in the spectra obtained for viscous media and glasses. They found that
the dispersion of the velocity of sound due to the relaxation of the shear viscosity could amount
to ~ 70%. This result was important also because some of the American scientists questioned
the existence of a fine structure of the Rayleigh line in the case of viscous liquids and glasses
[18].

Important results were also obtained by Shakhparonov and Tunin [19, 20], who carried out
further studies of the Rayleigh line.

No further studies of the fine structure of the Rayleigh line were made up to 1962. Although the earlier investigations produced important results and confirmed qualitatively the
relaxation theory in respect of low-viscosity liquids, a quantitative check of the theory could
not be carried out because it was practically impossible to determine simultaneously the velocity and absorption of hypersound. The principal difficulty was the excessive width of the
exciting line: the broadening of the Brillouin component represented only a small fraction of
the width of the exciting line.

Moreover, the exciting radiation provided by discharge lamps could not be used to find
details in the distribution of the intensity in the narrow part of the wing of the Rayleigh line.

The situation changed radically when investigators began to use gas lasers in laboratory
investigations. The present paper describes the first application of a gas laser in a study of
the position, profile, and width of the Brillouin components and of the structure of the narrow
part of the wing of the Rayleigh line in some liquids. The use of a gas laser made it possible
to determine the ratio of the integrated intensities of the components of the fine structure of
the Rayleigh line and to compare all the results obtained with the available theories. The results of these investigations are described in detail below.

The very first investigations of the width, profile, and positions of the Brillouin components, in which the present author participated [21, 22], made it possible to determine the amplitude coefficient of absorption of hypersound at ~ 5 × 10⁹ Hz and to measure simultaneously
the velocity of hypersound.

The use of a helium—neon gas laser as the exciting source enabled us to increase the
precision of the measurements of the velocity of hypersound by one order of magnitude, so
that this precision approached that attained in measurements of the velocity of ultrasound.
This higher precision made it possible to extend the range of substances in which dispersion
of the velocity of sound could be studied and to refine some of the earlier determinations of the
dispersion. The ratio of the integrated intensities of the components of the fine structure of
the Rayleigh line was determined with a precision much higher than in previous investigations
and it was possible to indicate which of the existing theories and formulas provided the best
description.

We were also able to observe a new phenomenon in the form of a fine structure in the narrow part of the wing of the Rayleigh line. This complex phenomenon was analyzed in detail and it was found that it could be used to determine the characteristics of the propagation of the transverse hypersound in low-viscosity liquids. Details of the apparatus with which these results were obtained are also given in the present paper.

Only a few years ago the Moscow physicists were the only ones investigating the fine structure of the Rayleigh scattering line. Since that time other investigators in the Soviet Union and abroad have followed suit. The flow of communications on the subject is now measured in tens of papers and it continues to accelerate, demonstrating the vitality of this new branch of physical optics.

CHAPTER I

SOME THEORETICAL AND EXPERIMENTAL INVESTIGATIONS OF THE MOLECULAR SCATTERING OF LIGHT IN LIQUIDS

§1. Fine Structure of the Rayleigh Scattering of Light in Liquids

The spectrum of the light scattered by fluctuations in liquids is a triplet consisting of two displaced Brillouin components (the Brillouin doublet) and the central undisplaced component. The Brillouin doublet appears because of the modulation of the scattered light by longitudinal hypersonic waves. Transverse (shear) hypersonic waves can also appear in liquids under some conditions. Modulation of the scattered light by the transverse waves gives rise to a doublet whose components are separated by a smaller interval than the components of the longitudinal doublet.

We shall now discuss the principal theoretical investigations concerning the scattering of light by fluctuations of the density and anisotropy in liquids.

We shall consider exciting radiation of wave vector $\mathbf{k}$ and assume that the scattered light is viewed along a wave vector $\mathbf{k'}$. Then, a maximum of the intensity of light scattered along the latter vector is observed if the vectors $\mathbf{k}$ and $\mathbf{k'}$ and the vector $\mathbf{q}$ of an elastic thermal wave satisfy the Bragg condition [2, 3]

$$\mathbf{k} \pm \mathbf{q} = \mathbf{k'}. \tag{1.1}$$

The frequency Ω_0 of the elastic thermal wave responsible for the scattering of light is given by [1]

$$\Omega_0 = 2n\omega_0 \frac{v}{c} \sin \theta/2, \tag{1.2}$$

where ω_0 is the frequency of the exciting monochromatic radiation; v is the velocity of the elastic thermal wave (the velocity of hypersound); c is the velocity of light; n is the refractive index; θ is the scattering angle. It follows from Eq. (1.2) that measurements of the frequency shift of the Brillouin components should give the phase velocity v of the elastic thermal wave of frequency $f = \Omega_0 /2\pi$. The frequency dependence of the velocity of hypersound v can be found by varying the scattering angle θ.

Fluctuations of the density, which result from the interference between elastic Debye waves, vary in time and space and modulate the scattered light. The modulation function $\Phi(t, r)$ can be found by solving the wave equation for the propagation of sound in a viscous medium (this can be done with the aid of the hydrodynamic theory [23]):

$$\ddot{\Phi} - v^2 \nabla^2 \Phi - \Gamma \nabla^2 \dot{\Phi} = 0, \tag{1.3}$$

where

$$\Gamma = \frac{1}{\rho} \left\{ \frac{4}{3} \eta + \eta' + \frac{\varkappa}{C_p} (\gamma - 1) \right\}. \tag{1.4}$$

Here, ρ is the density; $\varkappa$ is the thermal conductivity; C_p is the specific heat at constant pressure; $\gamma = C_p/C_v$; η and η' are the shear and bulk viscosities, respectively.

Equation (1.3) describes the propagation of an acoustic wave in a viscous medium.

Appropriate calculations [24, 25] give the spectral distribution of the intensity of light scattered by adiabatic fluctuations of the density:

$$I(\omega_0 \pm \Omega) = \frac{\Omega_0^2 \frac{2\delta}{\pi}}{(\Omega_0^2 - \Omega^2)^2 + 4\delta^2 \Omega^2}, \tag{1.5}$$

where $\delta = \Gamma q^2/2$; Ω is the angular frequency, measured from the position of the undisplaced line. A maximum of the intensity of a Brillouin component can be found from Eq. (1.5):

$$\Omega_{\max} = (\Omega_0^2 - 2\delta^2)^{1/2}. \tag{1.6}$$

If $\delta/\Omega \ll 1$, the half-width of a Brillouin component is

$$\delta\Omega \approx 2\delta \left(1 + \frac{1}{2} \frac{\delta^2}{\Omega^2} \right) \approx 2\delta = 2\alpha v, \tag{1.7}$$

where α is the amplitude coefficient of absorption of the elastic thermal wave. A convenient formula for the calculation of the half-width is

$$\delta\nu_B = \frac{\alpha v}{\pi c}, \text{ cm}^{-1} \tag{1.8}$$

It follows from Eq. (1.5) that the displaced (Brillouin) components should have a dispersion profile if the condition $\delta/\Omega \ll 1$ is satisfied. The thermal elastic wave is damped and, therefore, the position of the displaced component $\Omega_{\max}$ does not coincide with Ω_0, which is the frequency of this thermal wave. This should be borne in mind in measurements of the frequency shifts of the components and of the velocities of hypersound.

If a gas laser is used as the source of the exciting radiation, the velocity of hypersound can be measured to within 0.5-0.4% and the half-width of the shifted components $\delta\Omega$ can be measured to within a few percent of the displaced frequency Ω. In view of this high precision of the experimental results, an allowance must be made for the influence of the attenuation of hypersound (the width of the shifted components $\delta\Omega$) on the position of the Brillouin doublet.

If we use Eqs. (1.7) and (1.1), we find that [1, 26]

$$\delta\Omega_B = q^2 \Gamma = 2n^2 \Gamma |\mathbf{k}|^2 (1 - \cos\theta). \tag{1.9}$$

According to Eq. (1.9), the half-width of the Brillouin components depends strongly on the scattering angle θ and this must be allowed for in the determination of the width.

The half-width of the central component of the Rayleigh triplet, governed by isobaric fluctuations of the density, is given by the expression [1, 26]

$$\delta\Omega_c = 2n^2\chi k^2 (1 - \cos\theta), \tag{1.10}$$

where χ is the thermal diffusivity. It is evident from Eq. (1.10) that the central component is governed by the slow processes associated with the thermal diffusivity and that its half-width depends (like the half-width of the displaced components) on the scattering angle.

Estimates of the likely half-widths of the components of the fine structure of the Rayleigh line show that for most of the low-viscosity liquids it should be $\delta\nu_B \approx 10^{-2}$ cm^{-1} and the half-width of the central component should be $\delta\nu_c \sim 10^{-4}$ cm^{-1}.

The use of gas lasers enabled Fabelinskii et al. [21, 22] to determine the width of the Brillouin components for the first time. The application of the optical heterodyne method made it possible to determine also the width of the central component [27]. The results of these experiments indicate that the widths of the central and displaced components do indeed differ by almost two orders of magnitude.

Landau and Placzek [4] demonstrated that the ratio of the integrated intensities of the components in the Rayleigh triplet is given by the formula

$$I_c/2I_B = \gamma - 1, \tag{1.11}$$

where $\gamma = C_p/C_v$. Landau and Placzek gave Eq. (1.11) without derivation [4]. This derivation was later provided by several workers [28-30]. However, the quantity $(d\varepsilon/dT)_p k$ was ignored in [28-30], although its value could be quite considerable. Fabelinskii [6] included this quantity and showed that the ratio of the integrated intensities of the fine structure components should be

$$I_c/2I_B = L\frac{\sigma^2 T}{\rho C_p \beta_s}, \tag{1.12}$$

where $L = \left(\frac{1}{\sigma}\frac{\partial\varepsilon}{\partial T}\right)_p^2 \Big/ \left(\rho\frac{\partial\varepsilon}{\partial\rho}\right)_s^2$; σ is the volume expansion coefficient; ε is the permittivity; β_s is the adiabatic compressibility. Equation (1.12) reduces to Eq. (1.11) if the dispersion of the velocity of sound is ignored and the coefficient L is assumed to be unity.

The ratio of the integrated intensities of the fine structure components is frequently calculated from

$$I_c/2I_B = \frac{\delta\nu_c}{2\delta\nu_B}\frac{I_c^{max}}{I_B^{max}}, \tag{1.13}$$

where I_c^{max} and I_B^{max} represent, respectively, the maximum intensities of the central and Brillouin components. Equation (1.13) was suggested by Fabelinskii [6].

If we use Eqs. (1.2) and (1.7), we can derive the condition for resolution of the displaced components:

$$\alpha\Lambda \ll \pi, \tag{1.14}$$

where $\Lambda = \lambda/(2n \sin\theta/2)$ is the wavelength of the elastic thermal wave.

If the absorption coefficient of ultrasound is extrapolated quadratically to the frequencies of the Brillouin components ($f \sim 10^{10}$ Hz), it is found that the condition (1.14) is not satisfied ($\alpha \Lambda > \pi$) by benzene, carbon tetrachloride, and other liquids which exhibit a well-resolved fine structure of the Rayleigh line.

This contradiction was removed by Mandel'shtam and Leontovich [12], who developed a relaxation theory of the propagation of sound in condensed media. This theory will be considered in the next section.

§2. Relaxation Theory of the Propagation of Sound in Condensed Media

The relaxation theory of the propagation of sound developed by Mandel'shtam and Leontovich [12] is a general phenomenological theory in which no specific molecular model is used to describe the relaxation of the bulk viscosity.

This theory presupposes that the state of a liquid is described not only by its density ρ and the temperature T but also by a parameter ξ, which is a function of the density and temperature and which is $\xi = \xi_0$ under equilibrium conditions.

If we expand the rate of change of ξ with time as a series in terms of the difference $\xi - \xi_0$, we obtain [31]

$$\dot{\xi} = \frac{1}{\tau}(\xi - \xi_0). \tag{2.1}$$

The quantity τ has the dimensions of time and is known as the relaxation time. The process of re-establishment of equilibrium in a system slows down with increasing τ.

Let us assume that a liquid is subjected to periodic adiabatic compression and dilatation and that all the quantities which represent the state of the liquid have time dependences which include the factor $\exp(i\omega t)$. In this case, we find that $\xi_0 = \xi_{00} + \xi'$, where ξ_{00} is the constant component of ξ_0 and ξ' is the alternating component, proportional to $\exp(i\omega t)$. Similarly, the parameter ξ can be written in the form $\xi = \xi_{00} + \xi'$.

Calculations given in [31] show that the presence of slow processes in the re-establishment of equilibrium is macroscopically equivalent to the presence of the second (bulk) viscosity:

$$\eta' = \frac{\tau \rho}{1 + i\omega\tau}(v_\infty^2 - v_0^2). \tag{2.2}$$

If a slow process satisfies the inequality $\omega\tau \ll 1$, we find that

$$\eta_0' = \tau \rho (v_\infty^2 - v_0^2). \tag{2.3}$$

Substituting Eq. (2.3) into Eq. (2.2), we obtain an expression for the frequency dependence of the bulk viscosity:

$$\eta' = \frac{\eta_0'}{1 + i\omega\tau}. \tag{2.4}$$

It follows from the hydrodynamic theory that the amplitude coefficient of the absorption of sound is

$$\alpha = \alpha_\eta + \alpha_{\eta'} = \frac{\omega^2}{2v_0^3 \rho}\left(\frac{4}{3}\eta + \eta'\right). \tag{2.5}$$

If we use the real parts of Eqs. (2.3)–(2.5), we obtain

$$\alpha_{\eta'} = \frac{\omega^2}{2v_0^3 \rho}\,\frac{\eta_0'}{1 + \omega^2\tau^2} = \frac{\omega^2\tau\,(v_\infty^2 - v_0^2)}{2v_0^3\,(1 + \omega^2\tau^2)}. \tag{2.6}$$

It is evident from Eq. (2.6) that the component of the absorption due to the bulk viscosity relaxes to a constant value. This removes the contradiction between the conclusions of the hydrodynamic theory and the experimental results on the fine structure of the Rayleigh line.

Mandel'shtam and Leontovich [12] derived not only Eq. (2.6) but also an expression for the square of the complex velocity of sound:

$$u^2 = v_0^2 \left\{ 1 + \frac{i\omega\tau\,(v_\infty^2/v_0^2 - 1)}{1 + i\omega\tau} \right\}. \tag{2.7}$$

If we take the real part of the above expression, we obtain

$$\frac{v_\omega^2 - v_0^2}{v_\infty^2 - v_0^2} = \frac{\omega^2\tau^2}{1 + \omega^2\tau^2}. \tag{2.8}$$

It follows from Eqs. (2.7) and (2.8) that the velocity of sound should increase, tending to v_∞, when the frequency is increased. The relaxation time of the bulk viscosity, τ, is the only arbitrary parameter of the relaxation theory [12]. The determination of this relaxation time is a very important task. It can be carried out in two independent ways because τ occurs in the expression for the velocities (2.8) and in the expression for the absorption (2.6). If we solve Eqs. (2.3) and (2.8) for τ, we obtain

$$\tau = A + \sqrt{A^2 - \frac{1}{\omega^2}} \qquad \left(A = \frac{\eta'}{2\rho\,(v_\omega^2 - v_0^2)} \right). \tag{2.9}$$

Equation (2.9) can be used to determine the relaxation time τ by measuring the velocity of hypersound v_ω, the velocity of ultrasound v_0, and the bulk viscosity η'. If we measure that part of the width of the displaced components which is due to the bulk viscosity, we can find the corresponding absorption coefficient $\alpha_{\eta'} = \alpha - \alpha_\eta$ and, applying Eq. (2.6), we can determine τ by the second (independent) method.

If the two independent determinations of τ based on Eqs. (2.6) and (2.8) give the same value, we may conclude that the simple variant of the relaxation theory with a single relaxation time describes satisfactorily the propagation of sound in liquids.

If the relaxation process can be described by a single relaxation time, we can determine τ from the velocity of ultrasound v_0, the velocity of hypersound v_ω, and $\alpha_{\eta'}$. In this case, the expression for τ can be deduced from Eqs. (2.6) and (2.8):

$$\tau = \frac{v_\omega^2 - v_0^2}{2\alpha_{\eta'} v_0^3}. \tag{2.10}$$

The propagation of sound in real liquids usually has to be described by a set of relaxation times τ. Then, the value of τ deduced from Eq. (2.10) should be regarded as the average effective value.

Mandel'shtam and Leontovich [12] suggested a way for solving a more general problem in which the state of a liquid is described by several parameters ξ. However, in most cases it is not possible to distinguish the various relaxation processes in liquids which give rise to the dispersion of the velocity of sound and, consequently, the method suggested in [12] cannot be used.

The relaxation theory developed by Mandel'shtam and Leontovich is important because it is of a phenomenological nature although it fails to give specific information on the nature of the relaxation processes. This theory has provided a basis for experiments at ultrasonic and hypersonic frequencies in which the dispersion of the velocity of sound was observed for some liquids [1, 6, 19-20, 32].

The use of a gas laser in the present investigation made it possible to check directly the relaxation theory with a single relaxation time [12] against the results obtained for simple liquids. Surprisingly enough, it was found that the simple variant of the theory with a single relaxation time τ provided a correct quantitative description of the propagation of sound in some liquids. However, this simple theory was found to be inapplicable to other liquids.

§3. Phenomenological Theory of the Spectral Composition of Scattered Light and of the Propagation of Transverse Hypersonic Waves in Liquids

The spectral composition of light scattered by adiabatic and isobaric fluctuations of the density was discussed in §1. However, the influence of transverse hypersonic waves and of fluctuations of the anisotropy on the scattered-light spectrum was not considered.

A phenomenological theory of the spectral composition of the light scattered in liquids by fluctuations of the density and anisotropy, resulting from fluctuations of strain, was developed by Leontovich [5] and extended by Rytov [7].

The theory of Leontovich [5] is based on the Maxwellian treatment of the viscosity [33]. Leontovich [5] represents the state of a liquid by two tensors: the strain tensor e_{ik} and the anisotropy tensor ξ_{ik}. The derivatives of the strain tensor e_{ik} with respect to time are related to the components of the velocity:

$$\dot{e}_{ik} = \frac{1}{2}\left(\frac{\partial u_i}{\partial x_k} + \frac{\partial u_k}{\partial x_i}\right),$$

where the trace of the tensor $e_{\alpha\alpha} = \sigma$ is equal to the volume expansion coefficient. The tensor ξ_{ik} represents deviation of the liquid from isotropy and, therefore, $\xi_{\alpha\alpha} = \xi = 0$. Leontovich [5] points out that the description of the state of a liquid by just two tensors e_{ik} and ξ_{ik} restricts the validity of the theory.

If we consider small deviations from equilibrium and follow the calculations described in [5], we obtain the following three independent groups of equations of motion:

$$\rho \dot{u}_n - iqkE = 2\mu iq\Phi, \qquad \dot{\Phi} + \frac{1}{\tau}\Phi = \frac{2}{3}iqu_n; \qquad (3.1a)$$

$$\rho \dot{u}_l = 2\mu iqZ_{ln}, \qquad \dot{Z}_{ln} + \frac{1}{\tau}Z_{ln} = \frac{1}{2}iqu_l,$$

$$\rho \dot{u}_s = 2\mu iqZ_{sn}, \qquad \dot{Z}_{sn} + \frac{1}{\tau}Z_{sn} = \frac{1}{2}iqu_s; \qquad (3.1b)$$

$$\dot{X} + \frac{1}{\tau}x = 0, \qquad \dot{Z}_{st} + \frac{1}{\tau}Z_{st} = 0. \qquad (3.1c)$$

Here, the constants τ and μ are related by

$$\eta = \mu\tau. \qquad (3.2)$$

In Eq. (3.2) η is the usual shear viscosity, μ is the shear modulus, and τ is the anisotropy relaxation time. In the equations of motion (3.1) E is a component of the strain tensor; Z, Φ, and X are components of the anisotropy tensor; u are the components of the velocities.

The first group of equations, (3.1a), describes the propagation of the usual longitudinal waves responsible for the appearance of the Brillouin components. If we follow Rytov [7] and make an allowance for the temperature dependence, we find that the equations (3.1a) describe also the undisplaced line but, in this case, the adiabatic compressibility must be replaced with the isothermal value. These equations ignore the bulk viscosity which is the main factor that determines the width of the displaced component. It is evident from Eq. (3.1a) that the propagation of the longitudinal waves is accompanied by a change in the anisotropy provided $\Phi = Z_{nn} \neq 0$.

The second group of equations, (3.1b), describes the propagation of the transverse waves associated with the anisotropy of the liquid in question.

The last group of equations, (3.1c), represents fluctuations of the anisotropy which are not accompanied by the propagation of waves. Since, in this case, the only nonvanishing component of the tensor Z_{ts} is that which lies in the plane of the wave, we can speak of "transverse anisotropy waves," although no real waves are propagated.

We shall now calculate the distribution of the intensity in the scattered-light spectrum. We shall assume that light of frequency ω_0 is incident along the x axis and that it is linearly polarized along the l (z and y) axis. The field of the light wave scattered by the whole interaction volume is

$$E_p' = e^{i\omega_0 l} p_i \int\limits_V \Delta \varepsilon_{ik} e^{i(\mathbf{q}\mathbf{r})} \, dV.$$ (3.3)

The quantity $\Delta \varepsilon_{ik}$ can be written in the form [5]:

$$\Delta \varepsilon_{ik} = \left(\frac{\partial \varepsilon}{\partial \rho} \right) \Delta \rho \delta_{ik} + MG_{ik}.$$ (3.4)

If we substitute Eqs. (3.4) and (3.1) into Eq. (3.3), we obtain the electric field of the scattered wave. The intensity of the scattered light can be determined by finding the mean-square values of the coefficients in the Fourier transform of the scattered-wave field. Calculations for exciting radiation polarized linearly along the z axis yield:*

$$\begin{aligned}
I_{zz}(\omega_0 + \omega) &= \beta \left(\rho \frac{\partial \varepsilon}{\partial \rho} \right)^2 S_E(\omega) + \left(\rho \frac{\partial \varepsilon}{\partial \rho} \right)^2 A \left\{ \frac{1}{2} \beta S_{\Phi E}(\omega) + \right. \\
&\quad \left. + \frac{1}{6\mu} S_{E\Phi}(\omega) \right\} + A^2 \left\{ \frac{1}{2\mu} S_\Phi(\omega) + \frac{1}{4\mu} S_A(\omega) \right\}, \\
I_{yz}(\omega_0 + \omega) &= \frac{A^2}{8\mu} [S_T(\omega) + S_A(\omega)].
\end{aligned}$$ (3.5)

For light polarized linearly along the y axis we obtain

$$\begin{aligned}
I_{yz}(\omega_0 + \omega) &= \frac{A^2}{8\mu} [S_T(\omega) + S_A(\omega)], \\
I_{yx}(\omega_0 + \omega) &= \frac{A^2}{8\mu} \left[\frac{3}{2} S_\Phi(\omega) + \frac{1}{2} S_A(\omega) \right].
\end{aligned}$$ (3.6)

---

* The light polarized at right-angles to the scattering plane will be referred to as polarized along the z axis and the light polarized in the scattering plane will be described as polarized along the y axis.

If the interaction (scattering) region is illuminated with monochromatic unpolarized (natural) light we obtain the following expressions:

$$
\begin{aligned}
I_z(\omega_0 + \omega) &= \left(\rho\frac{\partial\varepsilon}{\partial\rho}\right)^2 \beta S_E(\omega) + \frac{1}{2}\left(\rho\frac{\partial\varepsilon}{\partial\rho}\right)^2 A \left\{\rho S_{\Phi E}(\omega) + \frac{1}{3\mu}S_{E\Phi}(\omega)\right\} + \\
&+ \frac{A^2}{4\mu}\left\{\frac{1}{3}S_\Phi(\omega) + \frac{1}{2}S_T(\omega) + \frac{3}{2}S_A(\omega)\right\}, \\
I_x(\omega_0 + \omega) &= \frac{A^2}{8\mu}\left\{\frac{3}{2}S_\Phi(\omega) + S_T(\omega) + \frac{3}{2}S_A(\omega)\right\}.
\end{aligned}
\tag{3.7}
$$

The following notation is used in Eqs. (3.5)–(3.7):

$$
\begin{aligned}
S_E(\omega) &= \frac{^8/_3\Omega_L^2\Omega_T^2\tau}{(\omega^2-\Omega_L^2)^2 + \omega^2\tau^2(\omega^2-\Omega_s^2)^2}, \qquad
S_\Phi(\omega) = \frac{3(\omega^2-\Omega_L^2)^2}{4\Omega_L^2\Omega_T^2}S_E(\omega), \\
S_T(\omega) &= \frac{2\omega^2\tau}{\omega^2+\tau^2(\omega^2-\Omega_T^2)^2}, \qquad
S_A(\omega) = \frac{2\tau}{1+\omega^2\tau^2}, \\
S_{\Phi E}(\omega) &= \frac{^4/_3\Omega_L^2(\omega^2-\Omega_L^2)\tau}{(\omega^2-\Omega_L^2)^2+\omega^2\tau^2(\omega^2-\Omega_s^2)^2}, \\
S_{E\Phi}(\omega) &= \frac{4\Omega_T^2(\omega^2-\Omega_L^2)\tau}{(\omega^2-\Omega_L^2)^2+\omega^2\tau^2(\omega^2-\Omega_s^2)^2}.
\end{aligned}
\tag{3.8}
$$

Here,

$$
\Omega_L = q\sqrt{\frac{1}{\beta\rho}} = \frac{4\pi v_L}{\lambda}\sin\theta/2, \qquad
\Omega_T = q\sqrt{\mu/\rho} = \sqrt{\frac{\eta}{\rho\tau}}\frac{4\pi}{\lambda}\sin\theta/2,
$$
$$
\Omega_s^2 = \Omega_L^2 + {}^4/_3\Omega_T^2.
\tag{3.9}
$$

It follows from Eqs. (3.7) and (3.8) that the distribution of intensities depends strongly on the frequency Ω_L of the longitudinal waves, the frequency Ω_T of the transverse waves, and on the anisotropy relaxation time τ; Ω_s is the frequency of the longitudinal waves in a solid whose shear modulus is μ. If $\Omega_L \ll \omega$ and $\Omega_T \ll \omega$, we find that

$$
S_\Phi(\omega) = S_T(\omega) = S_A(\omega) = \frac{2\tau}{1+\omega^2\tau^2},
$$

and then the expressions in Eq. (3.7) reduce to

$$
I_x(\omega_0 + \omega) = \frac{A^2 kT}{2\mu}\frac{2\tau}{1+\omega^2\tau^2}, \qquad
I_z(\omega_0 + \omega) = \frac{7A^2 kT}{12\mu}\frac{2\tau}{1+\omega^2\tau^2}.
\tag{3.10}
$$

The formulas in Eq. (3.10) describe the distribution of the intensity in the wing of the Rayleigh line and they show that, in the case of excitation with natural light, the depolarization coefficient of the wing is

$$
\frac{I_x}{I_z} = \rho(\omega) = \frac{6}{7},
\tag{3.11}
$$

However, if the incident monochromatic light is polarized linearly along the z axis, we find that $\rho(\omega) = \frac{3}{4}$, whereas, in the case of polarization along the y axis, we obtain $\rho(\omega) = 1$.

It follows from Eq. (3.5) that when the scattering is excited by light polarized along the z axis, the component of the scattered light polarized along the same axis should include the Brillouin doublet located in the shear wing. The component with the x-axis polarization should include the doublet due to the modulation of light by shear fluctuations of strain, i.e., due to the transverse elastic waves. In this case, the central line, due to the scattering of light by isobaric fluctuations of the density, and the Brillouin doublet are absent from the spectrum.

A comparison of Eqs. (3.5) and (3.6) shows that

$$I_{zx}(\omega) = I_{yz}(\omega).$$

Therefore, the component I_{yz} should include a doublet due to the transverse elastic waves.

Let us consider the spectral composition of the I_{yz} component of the scattered light. The term $S_A(\omega)$ of Eq. (3.8) describes the spectrum of the wing of the Rayleigh line. The term $S_\Phi(\omega)$ represents the "anisotropy waves" due to the longitudinal elastic waves. The spectrum of $S_\Phi(\omega)$ should include two maxima corresponding to $\omega = 0$ and $\omega = \Omega_s$ and a minimum corresponding to $\omega = \Omega_L$.

Leontovich does not use any specific molecular model of a liquid but gives a more general, phenomenological description. If we introduce specific models, such as the Debye model of a polar liquid, we find that [1]

$$\tau = \frac{1}{3}\tau_D = \frac{4}{3}\frac{\pi a^3 \eta}{kT}, \tag{3.12}$$

where τ_D is the relaxation time of the dipole moment in this liquid and a is the molecular size.

It must be stressed that the theory of Leontovich and Rytov [5, 7] is phenomenological and not restricted to any specific molecular model.

§4. Some of the Earlier Investigations of the Fine Structure and the Wing of the Rayleigh Line of Liquids

Fabelinskii [1, 13-15] was the first to determine reliably the dispersion of the velocity of sound from the fine structure of the Rayleigh line of liquids. Later, Fabelinskii and Pesin [16, 17] extended the range of the investigated substances and determined for the first time the fine structure of the Rayleigh line of viscous liquids such as glycerin and triacetin. Fabelinskii also considered the polarization of the fine structure components and obtained an expression (1.12) for the ratio of the integrated intensities of these components, which he compared with the experimental results.

Shakhparonov and Tunin [19, 20] also carried out some investigations of the fine structure of the Rayleigh line and extended further the range of liquids in which the dispersion of the velocity of sound was observed.

Fabelinskii [1, 6, 34] observed experimentally and investigated the narrow anisotropic wing of the Rayleigh line, which amounted to 0.1-1 cm^{-1} and was superimposed on the Brillouin components.* These narrow parts of the wing were explained easily by the rotational diffusion mechanism. The anisotropy relaxation times τ, deduced from the width of the narrow part of the wing, were in good agreement with Eq. (3.12) for all the liquids investigated. Similar results were obtained by Starunov [8].

All these investigations were carried out with the aid of mercury lamps which served as the exciting light sources. The half-width of the exciting light together with the instrumental

_____

*Fabelinskii investigated the temperature dependences of the half-width of the wing and its depolarization in phenyl salicylate (salol), triacetin, carbon disulfide, carbon dioxide, and benzene.

half-width was ~ 0.15-0.16 cm^{-1} (λ = 4350 Å) in all cases. Therefore, it was not possible to measure the width of the displaced components. Moreover, the use of such a wide exciting line meant that the velocity of hypersound could be determined only to within ~ 2-5% [1].

The use of lasers as the exciting light sources has lead to radical improvements in the experimental methods. The half-width of the exciting light provided by a gas (helium–neon) laser is an order of magnitude smaller (~ 0.01-0.02 cm^{-1}) than the corresponding half-width of the mercury line. This applies even in the multimode case when the instrumental half-width of the Fabry–Perot interferometer is included.

Mash, Starunov, Fabelinskii, and the present author [21, 22] were the first to use a helium–neon gas laser in investigations of the fine structure of the Rayleigh line. In these investigations we were able to determine for the first time the absorption coefficient of hypersound in liquids from the width of the Brillouin components and to improve by one order of magnitude the precision of determination of the velocity of hypersound.

These investigations were followed by further studies, mainly carried out in the USA [35-39], in which gas lasers were used. These studies yielded valuable data on the velocity and absorption of hypersound in liquids and on the dispersion of the velocity of sound.

However, it should be pointed out that Chiao and Fleury [40] failed to measure the width of the Brillouin components in nitrobenzene. We shall return to this point in the description of our measurements of the width of the same components.

We found that the velocity of hypersound deduced from the stimulated Brillouin scattering [41, 42] could not be regarded as absolutely reliable. We pointed out [42] that the velocity of hypersound deduced in this way was usually less than the value obtained with the aid of a gas laser. Some preliminary explanations of this effect were given in [43, 44] but the problem has not yet been solved. Therefore, the velocity of hypersound deduced from the stimulated Brillouin scattering can be used only in approximate estimates.

CHAPTER II

METHOD FOR EXPERIMENTAL INVESTIGATION OF THE THERMAL SCATTERING OF LIGHT IN LIQUIDS

§5. Apparatus Used in Investigations of the Fine Structure of the Rayleigh Line

The availability of gas lasers for laboratory investigations made it possible to modify greatly the method used in investigations of the spectra of the molecular scattering of light in liquids.

The apparatus used in the present investigation is shown schematically in Fig. 1. A cell V containing the scattering liquid was placed either in an external resonator, formed by mirrors M_1 and M_2, or in the laser resonator L formed by a window inclined at the Brewster angle Br and the mirror M_1. When placed inside the laser resonator, the cell V was oriented at the Brewster angle. In the external resonator a lens L_1 was used to increase the intensity of the exciting radiation. The focal length of this lens had to be sufficiently long ($f \approx$ 100-180 mm) because it determined the angle of convergence of the exciting beam (in the molecular scattering of light this beam should be nearly parallel).

The arrangement with the cell inside the laser resonator was preferable from all points of view. First of all, it ensured that the light beam was parallel. The intensity of the exciting

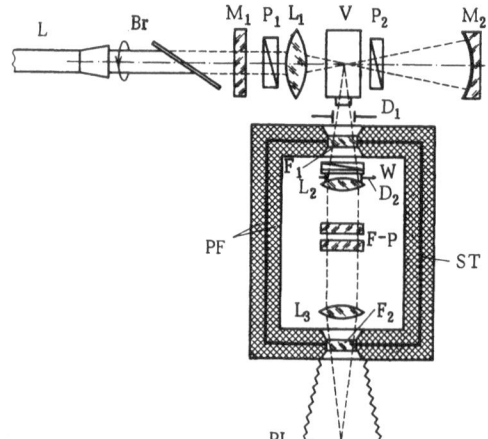

Fig. 1. Schematic diagram of the apparatus used. L is a laser; Br is a window aligned at the Brewster angle; M_1 and M_2 are mirrors; P_1 and P_2 are polarizers; L_1, L_2, and L_3 are lenses; V is a cell containing the liquid under investigation; D_1 and D_2 are diaphragms; F_1 and F_2 are filters; F-P is a Fabry—Perot interferometer; PF is a plastic foam jacket; ST is a steel casing; W is a Wollaston prism; PL is a photographic plate.

light was higher than in the external resonator. This was because the mirrors used in the gas laser had a transmission coefficient of ~1% and a reflection coefficient of ~98-99%. However, the arrangement with the cell inside the laser resonator was inconvenient when the exciting radiation was polarized in the scattering plane (the y polarization). In this case, it was necessary to use the external resonator. The lens L_1 focused the exciting beam inside the cell V and the spherical mirror M_2 ($f = 30$ cm) reflected and focused the transmitted light back into the cell [45]. The mirror M_2 had a multilayer dielectric coating and this was true also of the mirror M_1. Consequently, these two mirrors reflected a large fraction of the incident energy and increased the output power of the laser. Two polarizers (P_1 and P_2) were used to eliminate depolarization as a result of residual stresses in the glass walls of the cell V.

The scattered light was analyzed spectroscopically with a Fabry—Perot interferometer (F-P in Fig. 1). The interferometer mirrors had multilayer dielectric coatings of ~98% reflectivity. The resolution limit of the interferometer was ~0.002 cm^{-1} when the ring separating the mirrors was t = 10 mm thick.

A polarization analysis of the scattered-light spectrum was carried out with the aid of a Wollaston prism (W) which was placed in front of the collimator (L_2). The prism W was oriented so that one of its principal planes was perpendicular to the scattering plane (the z polarization) and the other was parallel to that plane (the x polarization). A quarter-wave plate was attached to the Wollaston prism to convert the linear polarization of light into the circular polarization and thus eliminate the influence of the difference between the conditions of propagation of the z and x polarized light.

The collimating objective L_2 directed a parallel beam of the scattered light onto the Fabry—Perot interferometer F-P and another objective L_3 was used to focus the spectrum onto a photographic plate PL. The exposure times were ~0.2-4 h, depending on the conditions.

In exposures lasting several hours the temperature and pressure of the ambient medium could vary significantly and this could distort the recorded spectral pattern [1]. The interferometer was protected from such temperature and pressure variations by a specially constructed isobaric and isothermal chamber. This chamber was a hermetically sealed welded steel box ST (the walls were ~10 mm thick) and the isothermal conditions were ensured by an outer layer of plastic foam PF (total thickness ~120 mm). Glass windows F_1 and F_2 were needed to ensure a hermetic seal. The chamber was tested for leaks and was found to ensure a satisfactory constancy of the pressure and temperature.*

---

*It was found that a vacuum of ~0.1 mm Hg was maintained for 10 h and the temperature remained stable to within ± 0.1°C.

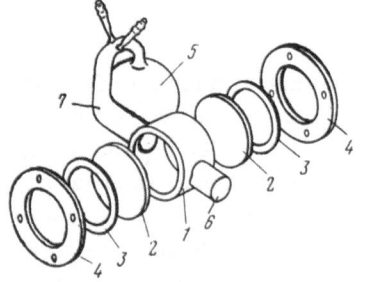

Fig. 2. Construction of the cell used in studies of scattering: 1) main part of the cell; 2) fused quartz windows; 3) rubber rings; 4) metal rings; 5) spherical reservoir; 6) window for scattered-light observations; 7) branch tube.

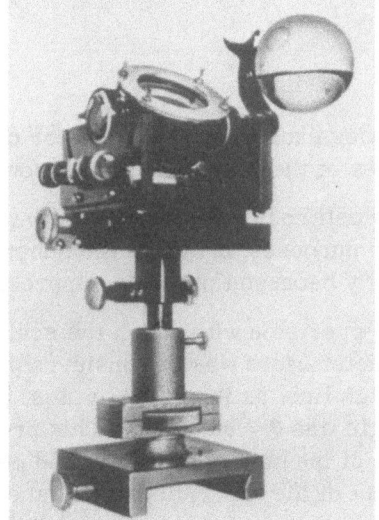

Fig. 3. External view of the cell in its holder.

The cell used was of special construction because of the need to insert it inside the laser resonator. In the earlier investigation [22] we used a cross-shaped cell with two windows oriented at the Brewster angle. Since then we developed a universal cell which was used in all our subsequent investigations [46]. The construction of the cell is shown schematically in Fig. 2 and its external appearance is pictured in Fig. 3.

The main part of the cell was in the form of a glass tube 1 with parallel ends. Two plane-parallel plates of optical-grade fused quartz were pressed against or bonded to (with a suitable adhesive) the ends of the tube. Thin (~ 1 mm) rubber rings were used as spacers and the whole assembly was bolted together with the aid of metal rings 4 (the bolts are not shown in Fig. 2). Two branch tubes 6 and 7 were attached to the main tube. One of these tubes (6) terminated in a plane-parallel window, used for the observation of the scattered light. The other tube (7) was horn-shaped and terminated in a spherical reservoir (5) of the liquid, which served as the "black" background in the observation of the scattered light. The two branch tubes were used to mount the cell in its holder and they served as a natural axis about which the cell could be rotated in the alignment at the Brewster angle. Purification of the liquid by the well-known Martin method [1] was carried out in the reservoir (5) attached to the branch tube (7). Thus, the main part of the cell was a plane-parallel container in which the liquid layer was ~ 30 mm thick.

The cell was filled with a purified liquid and aligned at the Brewster angle in the holder shown in Fig. 3. The holder was constructed in such a way that the cell could be oriented at the Brewster angle for any transparent liquid with a refractive index in the range from 1.30 to 1.70.

When the cell was placed in the laser resonator, it acted as a mode selector. The num-

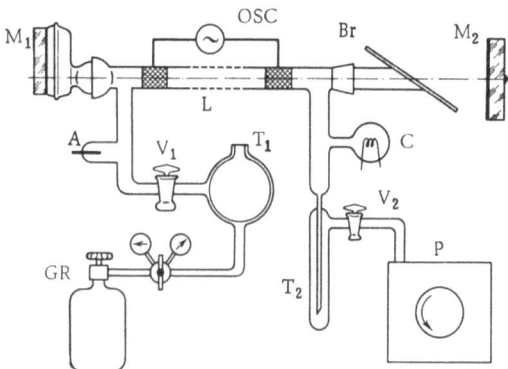

Fig. 4. Schematic diagram of the laser unit. L is a discharge tube; Br is a window aligned at the Brewster angle; M_1 and M_2 are mirrors; OSC is a high-frequency oscillator; V_1 and V_2 are vacuum valves; A is the anode; C is the cathode; T_1 and T_2 are traps; GR is a gas reservoir; P is a vacuum (backing) pump.

ber of modes excited in the resonator decreased with increasing difference between the refractive indices of the liquid and the window material.

The cell could also be used in investigations of the Raman scattering of light and for various other purposes, provided the substance placed in the cell did not absorb or scatter light too strongly because this could suppress the laser emission.

The precision with which the scattering angle and the aperture of the scattered light could be determined was of considerable importance in investigations of the fine structure of the Rayleigh line, as indicated by Eqs. (1.2) and (1.9). In the apparatus just described the scattering angle was $\theta = 90 \pm 0.2°$. This precision was achieved by the use of a pentaprism in the alignment of the optical system. The pentaprism rotated the laser beam through 90° and the optical part of the apparatus was aligned along this beam at right-angles to the original direction. The aperture of the scattered light was limited by two diaphragms (D_1 and D_2) and in all the experiments it amounted to ~ 0.03-0.04.

The apparatus described above had one additional advantage: the luminosity was sufficiently large for visual examination of the scattered-light spectrum. This was possible because of the use of a gas laser with a sufficiently high output power. Some of the spectra obtained with this apparatus are discussed in Chap. IV.

§6. Helium–Neon Gas Laser Used in Investigations of the Molecular Scattering of Light

We shall now describe the helium–neon gas laser constructed entirely in our laboratory and used in the excitation of the molecular scattering of light.*

The laser unit is shown schematically in Fig. 4. The laser discharge tube (L) was 15 mm in diameter and made of molybdenum glass with walls ~ 3 mm thick. One end of the tube terminated in a multilayer dielectric mirror M_1. This mirror reflected the radiation back into the tube and it was attached to a ball-and-socket joint. The other end of the tube terminated in a window (Br) oriented at the Brewster angle. The details of the two terminations of the laser tube can be seen in Figs. 5 and 6.

Zaitsev [47] suggested a method for attaching the laser mirrors and exit windows to conical and spherical joints. His suggestions were put into practice in the laser described here. The window Br inclined at the Brewster angle was attached to a conical joint. This made it possible to alter easily the polarization of the laser radiation, as required in our experiments.

---

*The author is grateful to V. P. Zaitsev for his great help in the construction of the laser.

Fig. 5. Mirror attached to the discharge tube by a ball-and-socket joint: 1) mirror; 2) conical joint whose axis is displaced relative to the laser axis; 3) ball-and-socket joint.

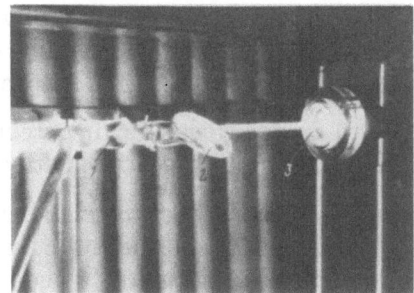

Fig. 6. Window aligned at the Brewster angle and exit mirror: 1) conical joint; 2) window aligned at the Brewster angle; 3) exit mirror.

The mirror M_1 was attached to a ball-and-socket joint (Fig. 4). The use of this joint enabled us to align the mirror quite easily without disturbing the pressure within the laser tube. The center of the mirror M_1 could be displaced relative to the axis of the tube and the whole mirror could be rotated in the joint. This arrangement made it easy to alter the position of the mirror M_1 in the optical resonator. The output power of this laser was higher, other conditions being equal, than that of a laser with two Brewster windows and external mirrors.

The laser unit was connected to a backing pump P. The vacuum that could be achieved in this way was $\sim 6 \times 10^{-3}$ mm Hg. The laser tube was filled with a mixture of helium and neon (in the 10:1 ratio) from a gas reservoir GR through a suitable regulator. The output power of the laser was ~ 20-25 mW at $\lambda = 6328$ Å.

§7. Method Used in the Analysis of Spectrograms and Estimated Precision of the Optical Measurements

A spectrogram of the scattered light can be used to measure the frequency shift of the Brillouin components and to determine the velocity of hypersound from Eq. (1.2). It can also be used to determine the width of the displaced components and to calculate the amplitude coefficient of the absorption of hypersound from Eq. (1.8). Moreover, the anisotropy relaxation time can be found from the half-width of the narrow part of the wing of the Rayleigh line.

An IZA-2 comparator was used in measurements of the linear separation of the fine structure components of the Rayleigh line. The frequency shift of the Brillouin components was calculated by the off-center method [1, 48]. The frequency shifts $\Delta \nu$ were substituted into Eq. (1.2) and the velocity of hypersound v was calculated at a given frequency. The relative error in the determination of the velocity of hypersound was the sum of the errors involved in the measurements of the quantities occurring in Eq. (1.2):

$$\frac{\Delta v}{v} = \frac{\Delta (\Delta v)}{\Delta v} + \frac{\Delta n^*}{n} + \frac{1}{2} \cot \theta/2 \, \Delta \theta. \qquad (7.1)$$

It is evident from Eq. (7.1) that the error in the determination of the velocity of hypersound depended primarily on the precision with which the frequency shift Δv and the scattering angle θ were measured.

The random error in the measurement of Δv was reduced by increasing the number of measurements of this shift in the same spectrum. The random error in the scattering angle θ was reduced by repeated alignment along this angle and photographing the spectrum each time. In the present study the scattering angle was set three times. When the values obtained in all three runs were averaged, the relative random error in the measurement of the velocity of hypersound was reduced to $\sim 0.1\text{-}0.5\%$.

The dispersion in an interferogram was known to decrease away from the center of the interference pattern. This effect was eliminated with the aid of a dispersion curve suggested by Fabelinskii [6]. The results of a correction applied in this way are shown in Fig. 7.

Corrected curves were used in determination of the widths of the Brillouin components and of the ratio of the integrated intensities of these components.

The profiles of the fine structure components were found to be dispersion curves and, consequently, the true width of the Brillouin components was determined by subtracting the instrumental width from the measured values. The instrumental width was assumed to be equal to the width of the central component. This approach was justified by the fact that the true width of the central component given by Eq. (1.10) was two orders of magnitude smaller than the width of the Brillouin components and, therefore, it could be ignored. The instrumental width of the central component was affected, like the widths of the displaced components, by the temperature and pressure variations during exposure.

The precision of measurement of the Brillouin components was $\sim 10\text{-}25\%$, depending on the experimental conditions. Since the fine structure components had dispersion profiles, the ratio of the integrated intensities could be calculated with the aid of Eq. (1.13).

All the above considerations were applicable to low-viscosity liquids such as benzene for which the wing of the Rayleigh line, superimposed on the fine structure components, was considerably wider than the separation between the Brillouin components ($\sim 1 \text{ cm}^{-1}$).

In the case of other liquids, part of the wing of the Rayleigh line superimposed in the region between the fine structure components had a half-width of $0.1\text{-}0.3 \text{ cm}^{-1}$. In such cases spectrograms were analyzed by a somewhat different method [49]. The scattered light was

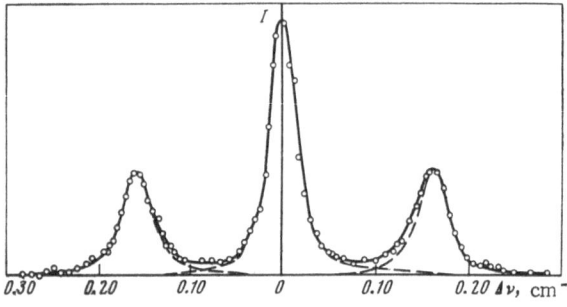

Fig. 7. Microphotogram of the spectrum of light scattered in benzene. The continuous curve represents the experimentally obtained intensity distribution and the dashed curve is the corrected distribution.

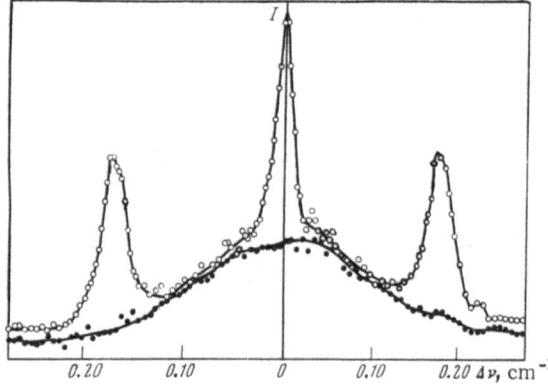

Fig. 8. Distribution of the intensity of light scattered in a liquid characterized by a narrow wing of the Rayleigh line (nitrobenzene). The open circles represent the z-polarization component and the black dots the x-polarization component.

Fig. 9. Results of correction of the intensity distribution shown in Fig. 8.

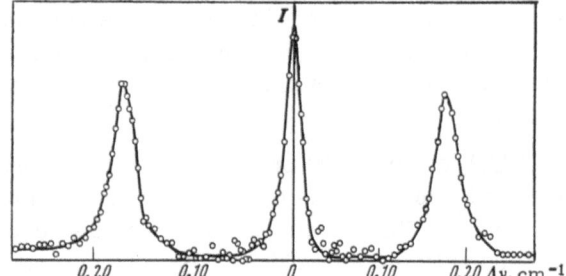

passed through a Wollaston prism and separated into the components with the z and x polarizations. In the spectrum with the z polarization the fine structure components were superimposed on the shear wing (Fig. 8), whereas in the spectrum with the x polarization only the wing of the Rayleigh line was observed. Therefore, the influence of the wing of the Rayleigh line was eliminated by subtracting the ordinate of the x component from the z component.* The results obtained in this way are plotted in Fig. 9.

CHAPTER III

DISPERSION OF THE VELOCITY OF SOUND AND THE ABSORPTION OF SOUND IN LIQUIDS

§8. Results of Measurements of the Velocity and Absorption of Hypersound in Liquids

The precision of measurement of the velocity of hypersound was improved considerably by the use of a gas laser in the investigations of the fine structure of the Rayleigh line. Consequently, it was possible not only to refine the earlier results on the dispersion of the velocity of sound but also to carry out new measurements on substances in which the dispersion could not be determined earlier. Moreover, the widths of the displaced components could now be measured and, consequently, the absorption coefficient of sound at frequencies of $\sim 10^9$-10^{10} Hz could be calculated.

---

* The ordinates of the x component were multiplied first by $\frac{4}{3}$ because the depolarization of the scattered light excited by the z-polarized radiation was $\frac{3}{4}$.

The velocity of hypersound was deduced from the formula

$$v = \frac{\Delta \nu c}{2 \nu n \sin \theta/2},$$ (8.1)

derived from Eq. (1.2). Here, ν is the frequency of the exciting radiation and $\Delta \nu$ is the separation between the displaced and the central components.

The velocity of hypersound was deduced from at least six spectrograms. This procedure reduced the random error in the measurements of the positions of the Brillouin components and of the scattering angle.

Different records of the fine structure were obtained for different temperatures of the scattering liquid. The temperature coefficient of the velocity of hypersound had been measured by various workers for acetone [14, 50] and dichloroethane [51]. These experiments showed that the temperature coefficients of the velocities of ultrasound and hypersound were identical. Therefore, the velocities of hypersound reported in the present paper were reduced to 20°C with the aid of the temperature coefficients of the velocities of ultrasound in the liquids under investigations.

Table 1 lists the principal physical properties of the liquids investigated. The refractive index at $\lambda = 6328$ Å was found by linear extrapolation of the known [52] values of n for $\lambda = 6563$ and 5893 Å. The bulk viscosity η' was deduced from ultrasonic measurements of the absorption of sound with the aid of Eq. (2.5). The velocities of ultrasound v_0 were taken from [32] and, in the case of water, acetone, and methylene bromide they were measured using apparatus described in [53]. Moreover, the temperature coefficient of the velocity of ultrasound in methylene bromide was measured and found to be $\Delta v/\Delta t = -2.8$ m·sec^{-1}·deg^{-1}.

We investigated ten liquids. The thermal Rayleigh scattering of light in benzene, carbon tetrachloride, toluene, chloroform, and methylene chloride and bromide had been investigated earlier by Fabelinskii et al. [1, 6, 13-15]. Similar measurements in quinoline were carried out by Shakhparonov and Tunin [19]. The fine structure of the thermal Rayleigh line of nitrobenzene, benzaldehyde, and aniline was investigated for the first time. The dispersion of the velocity of sound was refined for the liquids used in the earlier investigations and the coefficient of absorption of hypersound was determined for the first time from the width of the displaced components.

TABLE 1

Liquid	ρ, g/cm^3	n ($\lambda = 6328$ Å)	$\eta \cdot 10^3$, P	η', P	v_0, m/sec
Benzene	0.878	1.4993	6.5	0.921	1324
Carbon tetrachloride	1.595	1.4619	9.7	0.339	920
Methylene chloride	1.336	1.4140	4.3	1.071	1092
Toluene	0.867	1.4940	5.9	0.079	1324
Chloroform	1.498	1.4447	5.8	0.306	1000
Methylene bromide	2.495	1.539	1.03	0.654	961±1
Nitrobenzene	1.199	1.5483	20.3	0.116	1473
Water	0.997	1.3318	10.0	0.033	1486±1
Acetone	0.792	1.3575	3.16	0.015	1189±2
Benzaldehyde	1.05	1.5419	14.0	—	1479
Quinoline	1.095	1.6190	36.4	0.263	1600
Aniline	1.039	1.5819	44.3	—	1656

TABLE 2

Liquid	$f \cdot 10^{-9}$, Hz	v_h, m/sec	$\frac{\Delta v}{v} \cdot 10^2$	$\tau_D \cdot 10^{10}$, sec	$\alpha \cdot 10^{-3}$, cm^{-1}	v_∞, m/sec
Benzene	4.94 ± 0.02	1471 ± 8	10	2.5 ± 0.1	5.5 ± 0.4	1474 ± 8
Carbon tetrachloride	3.31 ± 0.01	1015 ± 6	10	0.93 ± 0.1	14.4 ± 2.3	1036 ± 13
Methylene chloride	(3.54) *	(1022)		(0.82)	(18.1)	(1043)
	3.531 ± 0.02	1113 ± 6	2	17 ± 5	0.91 ± 0.1	1113 ± 8
Toluene	4.53 ± 0.01	1357 ± 3	2.5	0.87 ± 0.1	3.80 ± 0.07	1363 ± 7
Chloroform	3.4 ± 0.01	1055 ± 5	5	1.68 ± 0.2	4.52 ± 1.0	1096 ± 14
Methylene bromide	3.37 ± 0.01	980.8 ± 1.4	2	7.07 ± 1.1	0.43 ± 0.12	981 ± 5
Nitrobenzene	5.31 ± 0.01	1535 ± 3 (1538)	4	0.26 ± 0.04	13.55 ± 2.24	1595 ± 18
Benzaldehyde	5.24 ± 0.01	1522 ± 4 (1525)	2.8	0.26 ± 0.08	10.5 ± 1.3	—
Quinoline	6.22 ± 0.01	1721 ± 2 (1728)	7.3	0.5 ± 0.01	19.1 ± 0.6	1758 ± 8
Aniline	6.06 ± 0.01	1714 ± 4	3.5	—	—	—

* The values in parentheses are obtained when an allowance is made for the influence of the absorption of hypersound on the positions of the Brillouin components.

TABLE 3

Liquid	$\delta v \cdot 10^3$, cm^{-1}	$\alpha \cdot 10^{-3}$, cm^{-1}	$\alpha_\eta \cdot 10^{-3}$	$\cdot 10^{-3}$ cm^{-1}	$\tau_\alpha \cdot 10^{10}$, sec	$\alpha/f^2 \cdot 10^{17}$
Benzene	7 ± 2	4.5 ± 1.3	2.0	2.5	3.03 ± 1.0	18.4 ± 5
Carbon tetrachloride						
Methylene chloride	17 ± 3	16 ± 3	2.0	14	0.88 ± 0.1	146 ± 27
	2 ± 2	—	—	—		
Toluene	10 ± 2	5.9 ± 1	1.58	5.36	0.49 ± 1	34 ± 7
Chloroform	11 ± 2	10 ± 2	1.15	9	0.96 ± 0.1	87 ± 17
Methylene bromide	4.9 ± 1.5	4.7 ± 1.5	0.14	4.58	1.73 ± 0.4	41 ± 13
Nitrobenzene	15 ± 2	9.2 ± 1.3	3.9	5.3	0.44 ± 0.1	32 ± 4
Benzaldehyde	17 ± 2	10.5 ± 1.3	3	7.5	0.26 ± 0.08	38 ± 5
Quinoline	30 ± 3	16.4 ± 1.6	8.2	8.2	0.54 ± 0.08	42 ± 4
Aniline	16 ± 2	8.8 ± 1.1	9.0	—	—	24 ± 3

Table 2 gives the velocities of hypersound (at 20°C) deduced from the shift of the Brillouin components and Table 3 lists the widths of these components. The relaxation times and other quantities calculated from the theoretical formulas (2.3), (2.6), (2.10), and (2.11) [12] on the assumption of a single relaxation time τ are also given in Tables 2 and 3.

The results given in Table 2 show that the precision in the measurements of the velocity of hypersound was ~ 0.1-0.5%. This precision set the limit to the smallest dispersion of the velocity of sound which could be measured. The widths of the Brillouin components were determined to within $\sim 2 \times 10^{-3}$ cm^{-1}.

The relative error in the measurement of the widths of the displaced components was still high compared with the error in the measurements of their positions. This was partly due to the photographic method of recording the spectrum because the graininess of the photographic material and its nonuniform development contributed to the experimental error. A somewhat higher precision could be achieved by photoelectric recording of the spectrum [40]. The high sensitivity of the photoelectric method made it possible to use a laser emitting a single mode and to increase the reflection coefficient of the interferometer mirrors. All these measures reduced the visible width of the central component and, consequently, the error in the measurements of the widths of the displaced components. However, it was difficult to record simultaneously the z and x polarized components of the scattered light when the Fabry—Perot interferometer was coupled to a photoelectric detector. Therefore, when this method was employed, the widths of the displaced components and the velocity of hypersound could not be measured accurately in liquids which had a narrow wing of the Rayleigh line (nitrobenzene, quinoline, etc.). This was why Chiao and Fleury [40] were unable to determine the widths of the displaced components of the scattering in nitrobenzene.

In our earlier experiments [54] we analyzed the polarization of the scattered light and this enabled us to eliminate, at least partially, the influence of the depolarized part of the scattered light on the positions of the Brillouin components.

§9. Discussion of Experimental Results on the Propagation of Hypersound in Liquids

Simultaneous and independent measurements of the velocity of hypersound and of the widths of the Brillouin components enabled us to check directly the simplest variant of the relaxation theory with a single relaxation time [12].

Tables 2 and 3 give the experimental values of the widths of the displaced components and of the velocities of hypersound alongside the results of calculations of the absorption coefficient of hypersound α, the relaxation time of the bulk viscosity τ, and the limiting (high-frequency) value of the velocity of hypersound v_∞ deduced from Eqs. (2.6)-(2.11).

A comparison of the relaxation times of the bulk viscosity calculated from the dispersion of the velocity of sound (τ_D) and from the widths of the displaced components (τ_α) demonstrated that the simple variant of the relaxation theory with a single relaxation time described satisfactorily the properties of benzene and carbon tetrachloride. This conclusion was drawn from the fact that the relaxation times τ_D and τ_α were equal (within the limits of the experimental error) for these liquids.

In some cases, it was possible to determine the extent to which the simple relaxation theory (with a single value of τ) was satisfied by plotting the dependences of α/f^2 on $\log f$. According to the relaxation theory with a single τ [12], we should have

$$\alpha/f^2 = B + \frac{A^2}{(1 + f^2/f_c^2)} , \qquad (9.1)$$

where

$$B = \alpha\eta/f^2 = \frac{8\pi^2\eta}{3v_0^3\rho} , \qquad A = \alpha\eta'/f^2 = \frac{2\pi^2\eta'}{v_0^3\rho} ,$$

$f_c = \frac{1}{2}\pi\tau$, and τ is the relaxation time of the bulk viscosity η'. We used Eq. (9.1), the measured values of the absorption coefficient of ultrasound [32, 55-59], and our own measurements at hypersonic frequencies to plot the dependences of α/f^2 on $\log f$ for benzene and carbon tetrachloride. A dependence of this type obtained for benzene (Fig. 10) indicated that the measurements of different workers could be described satisfactorily by the simple theory. This agreement between different measurements and the theoretical predictions confirmed the existence of a process with a single relaxation time in benzene at $f \sim 5 \times 10^9$ Hz. Therefore, it is difficult to understand why Hunter et al. [58] concluded that the relaxation process in benzene is complex and has several relaxation times τ. The results obtained by Hunter et al. [58] in ultrasonic measurements of the absorption are in good agreement with our data (Fig. 10). Moreover, the

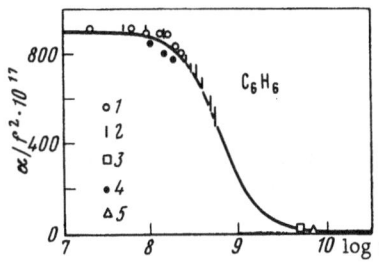

Fig. 10. Dependence of α/f^2 on log f for benzene.
1) [59]; 2) [58]; 3) our result; 4) [57]; 5) [40].

Fig. 11. Dependence of α/f^2 on log f for nitrobenzene. The open circles represent results
reported in [57]; 1), 2) our data.

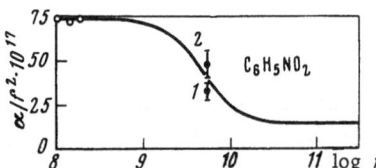

validity of the simple variant of the relaxation theory [12] in the case of benzene is fully supported by hypersonic measurements [40] carried out at higher frequencies. A comparison of
the experimental and theoretical values [1] shows that the simple relaxation theory describes
also the behavior of carbon tetrachloride and carbon disulfide.

The dispersion of the velocity of sound in nitrobenzene was first discovered by us in an
investigation of the stimulated Brillouin scattering [42]. We later carried out a more detailed
and precise study of the relaxation of the bulk viscosity of nitrobenzene [54]. We found that
this relaxation process can be described by a single relaxation time since the values of τ deduced from the dispersion and absorption of sound are very similar (Tables 2 and 3). On this
basis we plotted the dependence of α/f^2 on log f for nitrobenzene (Fig. 11). The results of
direct measurements of the dispersion of the velocity of sound in nitrobenzene are in agreement (within the limits of the experimental error) with our measurements based on the stimulated Brillouin scattering [42]. The disagreement between our values of the velocity of hypersound and the results reported by Chiao and Fleury [40] is due to deficiencies of the experimental method used in [40]. Chiao and Fleury did not separate the scattered light into components with mutually perpendicular polarizations. Therefore, Chiao and Fleury failed to eliminate the influence of the depolarized part of the scattered light, which is very significant in
the case of benzene.

Shakhparonov and Tunin [19] were the first to observe the dispersion of the velocity of
sound in quinoline (at 70°C) and to calculate the relaxation parameters of this liquid. These
measurements could not be carried out at 20°C because of the strong fluorescence exhibited
by quinoline when it was excited with the mercury line at $\lambda = 4358$ Å. The use of a helium—neon
gas laser ($\lambda = 6328$ Å) allowed us to avoid this influence of fluorescence and to carry out measurements at 20°C. Moreover, the laser light enabled us to measure the width of the displaced
components for quinoline, which was found to be largest ($\delta\nu = 30 \times 10^{-3}$ cm^{-1}) among all the
liquids studied so far.

A comparison of the relaxation times calculated from the dispersion of the velocity of
sound (τ_D) and from the absorption of sound (τ_α) demonstrated that the behavior of quinoline
can be described by the simple variant of the relaxation theory with a single value of τ. This
is supported by the dependence of α/f^2 on log f (Fig. 12). Thus, five liquids (benzene, carbon
tetrachloride, carbon disulfide, nitrobenzene, and quinoline) obey the simple variant of the relaxation theory with a single value of τ.

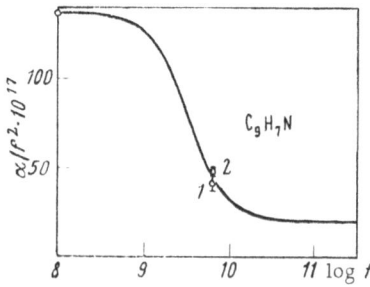

Fig. 12. Dependence of α/f^2 on log f for quino-line. The open circle located at low frequencies is taken from [20]; 1), 2) our data.

Fig. 13. Dependence of α/f^2 on log f for methyl-ene bromide. The vertical dashes are the results taken from [64]; 1) value calculated from the dispersion of the velocity of sound; 2) value cal-culated from the width of the displaced compo-nents.

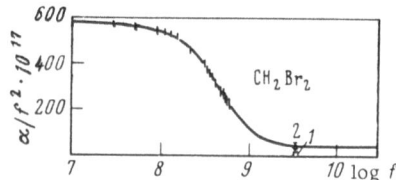

The first hypersonic study of methylene chloride, methylene bromide, and chloroform was carried out by Pesin and Fabelinskii [15, 60]. Earlier studies of the absorption of ultra-sound in vapors [61, 62] showed that methylene chloride and bromide have two relaxation re-gions.

Our measurements of the velocity of hypersound made it possible to refine the data on the dispersion of the velocity of sound obtained for these liquids by Pesin and Fabelinskii [15, 60]. The values of the absorption coefficient of hypersound α obtained experimentally for methylene bromide and chloroform were used to compare the relaxation times τ deduced separately from the dispersion of the velocity of sound and from the absorption of sound. Such a comparison (Tables 2 and 3) demonstrated that the relaxation in chloroform and methylene bromide is not described by a single relaxation time because τ_D and τ_α are not equal. However, this does not mean that two relaxation processes occur in these liquids and, moreover, it is not possible to separate experimentally different relaxation processes. The dependence α/f^2 on log f plotted in Fig. 13 for methylene bromide is adjusted to fit the ultrasonic measurements [54]. This fig-ure includes also our data on the absorption deduced from the width of the displaced compo-nents and from the dispersion of the velocity of sound. The ultrasonic measurements of the absorption [63] and our values of the coefficient of absorption of hypersound α, deduced from the width of the Brillouin components, are described satisfactorily by the dependence of α/f^2 on log f plotted using the ultrasonic data [63]. This dependence suggests the existence of sev-eral relaxation times because the value of B obtained experimentally is higher than the value of $\alpha\eta/\alpha^2$ given by Eq. (9.1). Consequently, the relaxation of the bulk viscosity in methylene bro-mide is not yet complete at $\sim 3 \times 10^9$ Hz and should be observed at higher frequencies. It is worth noting that when several relaxation processes take place in a liquid, the absorption co-efficient calculated from the dispersion of the velocity of sound may differ from the coefficient deduced directly from the width of the Brillouin components (Fig. 13) because the calculations are made using theoretical formulas derived on the assumption of a single relaxation time of the bulk viscosity.

The experimental results for methylene chloride were difficult to interpret because the width of the displaced components was close to the experimental error: $(2 \pm 2) \times 10^{-3}$ cm^{-1}. Hanes et al. [64] recently carried out a photoelectric spectroscopic study and determined the absorption of hypersound in methylene chloride from the width of the displaced components and obtained $(4 \pm 1) \times 10^{-3}$ cm^{-1}. If this value is accurate, we must assume that several re-

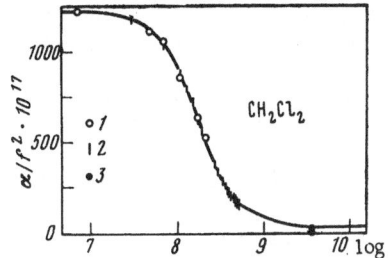

Fig. 14. Dependence of α/f^2 on log f for methylene chloride. 1) [72]; 2) [63]; 3) [65]; 4) our value deduced from the dispersion of the velocity of sound.

Fig. 15. Dependence of α/f^2 on log f for toluene. The open circles represent results taken from [57]; 1), 2) our data.

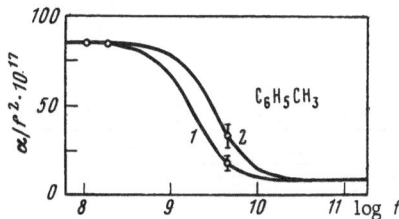

laxation processes occur in methylene chloride. The dependence of α/f^2 on log f (Fig. 14) plotted using the results of ultrasonic measurements [65] demonstrates good agreement between the measured absorption of hypersound at $\sim 3.5 \times 10^9$ Hz with the curve plotted on the assumption that several relaxation processes occur in methylene chloride. The absorption coefficient of hypersound, calculated from the measured dispersion of the velocity of sound, is also plotted in Fig. 14. Methylene chloride exhibits, like methylene bromide, a complex relaxation process with several relaxation times τ (B > α_η/f^2) and the value of τ given in Table 2 should be regarded as approximate.

The data on toluene, listed in Tables 2 and 3, demonstrate that the relaxation times τ deduced from the dispersion of the velocity of sound, do not agree with the corresponding values deduced from the absorption of sound. Although this disagreement is not as large as in the case of methylene bromide, nevertheless, we must assume the occurrence of several relaxation processes in toluene. The dependence of α/f^2 on log f for toluene (Fig. 15) supports this conclusion. Moreover, our results are in full agreement with those given in [40, 66].

The fine structure of the thermal scattering of light in benzaldehyde and aniline was investigated for the first time. Measurements of the absorption and dispersion of the velocity of sound in benzaldehyde (Tables 2 and 3) indicated the occurrence of relaxation of the bulk viscosity. However, calculations based on the relaxation theory [12] were difficult because of the absence of data on the absorption of ultrasound. Therefore, the relaxation time of the bulk viscosity τ of benzaldehyde was calculated from Eq. (2.11) and included in Tables 2 and 3. It should be stressed that the relaxation time found from Eq. (2.11) is approximate because it presupposes that the bulk viscosity has a single relaxation time τ. Our experience shows that this simple variant of the relaxation theory [12] does not apply to many liquids. Therefore, the relaxation time calculated for benzaldehyde should be regarded as an order-of-magnitude estimate.

Garmire and Townes [41] used the stimulated Brillouin scattering to determine the velocity of hypersound in aniline, which was in agreement with our measurements (Table 2). Our absorption coefficient α of hypersound in aniline (Table 3) agreed with the absorption coefficient of sound due to the shear viscosity α_η. Since aniline exhibited dispersion of the velocity of sound, this agreement between the absorption coefficient α and α_η indicated that the bulk viscosity of the aniline was fully relaxed at $f \sim 6 \times 10^9$ Hz. Therefore, studies of the relaxation of the bulk viscosity of aniline should be carried out at lower frequencies ($\sim 10^7$-10^8 Hz).

The relaxation parameters of the liquids discussed above could not be calculated more precisely because of the lack of data on the absorption of ultrasound in these liquids.

Summarizing the results obtained, we can say that some of the liquids can be described by the simple relaxation theory with a single relaxation time (this applies to benzene, carbon tetrachloride, carbon disulfide, nitrobenzene, and quinoline), whereas more complex relaxation processes occur in other liquids such as toluene, chloroform, and methylene chloride and bromide. Since all the calculations were carried out using the formulas of the relaxation theory presupposing a single value of τ, the relaxation times computed for toluene, chloroform, and methylene chloride and bromide (Tables 2 and 3) should be regarded as approximate.

§10. Velocity of Hypersound and Absorption of Sound in Water and Acetone. Negative Dispersion of the Velocity of Sound

In the preceding sections we have considered liquids which obey the simple relaxation theory of [12] but exhibit dispersion of the velocity of sound and deviation from the quadratic frequency dependence of the absorption coefficient α. These effects are manifestations of relaxation processes. It should be noted that all the liquids discussed above exhibit a positive dispersion of the velocity of sound ($\partial v/\partial f > 0$), i.e., the velocity increases with increasing frequency. We shall now consider the possibility of the existence of a negative dispersion ($\partial v/\partial f < 0$).

The first investigation of the dispersion of the velocity of sound [67] demonstrated the existence of a negative dispersion of $\sim 20\%$ in acetone. This negative dispersion could not be explained by the simple relaxation theory [12] because the relaxation of the bulk and shear viscosity [33] should give rise to a positive dispersion. The negative dispersion was attributed in [68] to the adiabatic–isothermal transition at the hypersonic frequency. Ginzburg [69] showed that the explanation proposed in [68] was untenable under the conditions in question. The frequency corresponding to the transition from isothermal to adiabatic conditions in the liquids in question is $\sim 10^{12}$-10^{13} Hz [1], whereas the frequency involved in the scattering of light is $\sim 4 \times 10^9$ Hz.

Ginzburg [69] and Vladimirskii [28] demonstrated that, in principle, the velocity of sound can have a negative dispersion. Both theories deal with the dispersion which is not related specifically to the bulk or shear viscosities. Vladimirskii [28] gives a generalized thermodynamic calculation of the dispersion of the velocity of sound, whereas Ginzburg [69] discusses the Navier–Stokes equation in the second approximation. The conclusions of both theories can be written in the form

$$\frac{\Delta v}{v} = \frac{q^2}{2v^2} f^*, \tag{10.1}$$

or in a form more convenient in calculations,

$$\frac{\Delta v}{v} \sim 2\pi^2 \left(\frac{r}{\Lambda}\right)^2, \tag{10.2}$$

where q is the wave number of sound and f^* is the intermolecular interaction constant. Equation (10.2) is obtained from Eq. (10.1) by assuming that $|f^*| \sim v_0^2 r^2$, where r is the correlation radius or the size of a complex.

TABLE 4

Liquid	t, °C	$f \cdot 10^{-9}$, Hz	v_h, m/sec	v_u, m/sec	$\frac{\Delta v}{v} \cdot 10^2$	Reference
Water	20	6.4	1480±20	1485	0	[1]
	22.9	4.33	1457±10	1492	—2.4	[35]
	20	2.951	1490±20	1484.5	0	[73]
	22	5.66	1471±8	1490	—1.28	[41]
	20	6.15	1470±7	1484.5	—1.0	[55]
	27.7	6.31	1504±1.4	1503.5	0	[37]
	24.7	5.66	1466±37	1492	—1.7	[70]
	22.0	6.24	1488	1490	0	[40]
	26.6	4.47	1468±10	1501	—2.2	[71]
	21.4		1476	1488.5	—0.8	[66]
Acetone	20	5.3	1190±40	1192	0	[6]
	22	3.47	1144±6.5	1181	—3.13	[35]
	22	4.59	1174±7	1181	—0.6	[41]
	20	2.951	1180±20	1192	0	[73]
	22		1180	1181	0.	[66]
	21.6	5.05	1190	1183	0	[40]
	27.4	3.511	1136.8±3.0	1150	—1.15	[71]

TABLE 5

Liquid	t, °C	Δv, cm^{-1}	$f \cdot 10^{-9}$, Hz	v_h, m/sec	v_u, m/sec	$\delta v \cdot 10^3$, cm^{-1}
Water	20	0.1474±0.0002	4.42	1486±1.3	1486±1.0	6±1.6
Acetone	20	0.1202±0.0003	3.60	1190±2.5	1190±2	4±1.6

The sign of f^*, which depends strongly on the intermolecular interaction, determines the sign of the dispersion of the velocity of sound. If we use the model of hard spheres which do not interact elastically, we find that $f^* > 0$. In a system of molecular complexes bound by elastic forces we can have $f^* < 0$ and the dispersion of the velocity of sound can be negative. A liquid can be regarded as a hard-sphere medium only in the vicinity of the critical point and, therefore, it is only in this case that we have $f^* \geq 0$. In all other cases, we should obtain $f^* < 0$ and, consequently, the dispersion of the velocity of sound should be negative. Estimates show that this dispersion should be weak. For example, under the conditions employed in [67], we find that $\Delta v/v \sim 0.05\%$ if r $\sim 10^{-7}$ cm far from the critical region. A similar calculation for water yields $\Delta v/v \sim 0.02\%$. If the negative dispersion is $\sim 20\%$, the correlation radius should be $\sim 10^{-6}$ cm, which is unlikely to be correct under normal conditions. Moreover, later investigations [6] failed to confirm the results obtained by Rao [67].

Recent investigations carried out using gas lasers [35–37, 40, 41, 66, 70–72] and ultrasonic measurements carried out at $f = 2.9 \times 10^9$ Hz [73] made it desirable to carry out a serious experimental study of the negative dispersion of the velocity of sound. The experimental results published by various workers are collected in Table 4. It is evident from this table that the negative dispersion found by some workers exceeds the experimental error. Unfortunately, the velocity of ultrasound was not measured in the investigations of the negative dispersion of the velocity of hypersound in water and acetone. Such measurements are essential in the search for small dispersions of the velocity of sound.

The results discussed in the present section show clearly that the question of the existence of the negative dispersion of the velocity of sound is important from the point of view of

TABLE 6

Liquid	$f \cdot 10^9$, Hz	$\alpha/f^2 \cdot 10^{17}$, cm^{-1}Hz^{-2}	Reference
Water	0.25	25	[32]
	0.19	21.2	[57]
	4.4	19$\pm$5	[74]
Acetone	0.4	25.7	[57]
	3.6	25$\pm$10	Our data and [74]

molecular physics. Therefore, we determined the velocity of hypersound and ultrasound and measured the width of the Brillouin components in the same samples of water and acetone [74]. The experimental results obtained in this way are given in Table 5. The velocities of hypersound and ultrasound were measured to within ~0.1-0.2%. The results indicated that neither positive nor negative dispersion was observed for water and acetone. The velocity of hypersound in water at $f = 6.31 \times 10^9$ Hz [37] also showed no dispersion.

The measured values of the absorption coefficient of hypersound α were used to calculate α/f^2 for water and acetone. These hypersonic values were compared with the values of the same ratio deduced from ultrasonic measurements [32, 57]. The results of this comparison (Table 6) showed that the relaxation of the bulk viscosity was not observed at $f \sim 4 \times 10^9$ Hz. Our measurements of the absorption were confirmed by a study of the absorption of sound in water at $f = (0.5$-$1.5) \times 10^9$ Hz reported in [75], which also failed to reveal any relaxation of the bulk viscosity. The precision of the measurements described in [75] was $\pm$ 1%.

The experimental data on the absorption of sound in water are plotted in Fig. 16 in the form of the dependence of the ratio α/f^2 on log f. The results of measurements of different workers [40, 74, 75] confirmed the absence of relaxation of the bulk viscosity in water right up to ~6.3×10^9 Hz. Measurements of the absorption reported in [37] should be regarded as inaccurate because they indicated relaxation of the bulk viscosity in water at ~6.3×10^9 Hz. If such relaxation were to occur at this frequency, the dispersion of the velocity of hypersound, studied in [37], should be ~2%. However, no dispersion was observed in [37]. Our results are also in agreement with the theoretical conclusions [28, 69] provided the linear dimensions of molecular complexes in the liquids under investigation were less than 10^{-7} cm. This assumption was quite reasonable and it was in agreement with the theory of intermolecular interactions.

We shall now consider the conclusion that can be drawn from a comparison of our experimental results and the theoretical papers of Hall [76] and Eucken [77], dealing with the structural relaxation in water.

Hall [76] assumes that water molecules can form two structures whose presence is a function of temperature. One of these structures resembles the lattice of ice and the other is more closely packed and its specific weight increases with rising temperature.

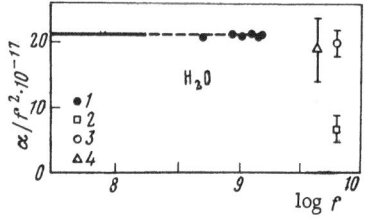

Fig. 16. Dependence of α/f^2 on log f for water. 1) [75]; 2) [37]; 3) [40]; 4) our data.

Eucken [77] suggests that water consists of single H_2O molecules as well as double $(H_2O)_2$, quadruple $(H_2O)_4$, and octuple $(H_2O)_8$ molecules. An octuple molecule, $(H_2O)_8$, occupies a larger volume than eight single molecules, etc. This increase in volume is responsible for the increase in the specific heat of water.

The estimates given in [78] yield a relaxation time $\tau = 2 \times 10^{-12}$ sec and a characteristic relaxation frequency $f_c \sim 7 \times 10^{10}$ Hz for the Hall theory [76]. The corresponding values for the Eucken theory are $\tau = 4 \times 10^{-11}$ sec and $f_c \sim 4 \times 10^9$ Hz. Our results on the absorption and dispersion of the velocity of sound are incompatible with the Eucken theory because a positive dispersion of $\sim 1.5\%$ follows from the Eucken theory. The Hall theory is not in conflict with our measurements of the dispersion of the velocity and the absorption of sound but this is insufficient argument in support of this theory.

§11. Ratio of the Integrated Intensities of the Fine Structure Components of the Rayleigh Line

The high resolution of the scattered-light spectra obtained with the aid of a helium—neon laser ($\lambda = 6328\,\text{Å}$) enabled us to determine the ratio of the integrated intensities of the fine structure components (the Landau—Placzek ratio) more accurately than in earlier studies [1, 6]. The profiles of all the components of the fine structure [22] were determined for the first time and more accurate measurements were carried out.

The ratio of the integrated intensities was found by the method of maximum intensities. Graphs similar to that shown in Fig. 7 were employed in the determination of the maximum intensities of the components and of their visible half-width. The ratio of the integrated intensities was deduced from Eq. (1.13). This method could be employed because the visible profiles of the fine structure components of the Rayleigh line could be described by dispersion curves.

The spectra of the light scattered in low-viscosity liquids (benzene, toluene, etc.), which did not exhibit a narrow wing of the Rayleigh line, were not photographed in two mutually perpendicular polarizations because the width of the narrowest part of the wing of the Rayleigh line was ~ 2 cm^{-1}, whereas the spectral range of the interferometer was ~ 0.5-0.6 cm^{-1}. Therefore, we assumed that the distribution of intensity in the continuous spectrum was uniform.

The spectra of light scattered in liquids exhibiting a strong narrow wing (nitrobenzene, quinoline, etc.) were separated into components with the z and x polarizations and interpreted by the method described in Chap. II.

The values of the Landau—Placzek ratio obtained in this way are listed in Table 7. This table also includes the values obtained recently by other workers and the values calculated from Eqs. (1.11) and (1.12).

It is easiest to compare the results of measurements with the calculations based on Eq. (1.11). Such a comparison (Table 7) shows that the agreement with Eq. (1.11) is unsatisfactory for most of the liquids under investigation. A complete agreement is obtained only for acetone, in which no dispersion of the velocity of sound is observed. The greatest discrepancies occur for those liquids in which the dispersion of the velocity of sound is strongest. This experimental observation forms the basis of the theory developed by Fabelinskii [6] for the ratio of the integrated intensities of the fine structure components of the Rayleigh scattering of light in liquids. An allowance for the dispersion of the velocity of sound led Fabelinskii to Eq. (1.12), which can be reduced to Eq. (1.11) if certain assumptions are made (see §1).

A comparison of the measured values of the ratio of the integrated intensities, $I_c/2I_B$, with the results of calculations based on Eq. (1.12) is difficult because many of the quantities which occur in Eq. (1.12) are either not available at all or are known inaccurately. It is shown

TABLE 7

Liquid	Measured values	Calculated using		Reference
		(1.12)	(1.11)	
Benzene	0.72±0.1	0.60	0.46	Our data
	0.84±0.05		0.43	[80]
	0,83			[79]
Carbon tetrachloride	0.67±0.15	0.56	0.46	Our data
	0.72±0.03		0.45	[80]
	0.68	0.56		[79]
	0.75		0.48	[71]
Chloroform	0.58±0.1	0.58	0.51	Our data
Methylene chloride	0.72±0.1	—	0.54	Ditto
Methylene bromide	0,54±0.05	—	0.47	Ditto
Acetone	0.44±0.01	0.47	0.39	[80]
	0.40±0.05		0.39—0.40	Our data
	0.44		0.42	[71]
Toluene	0.45±0.06	0.43	0.36	Our data
	0.42±0.03	0.41	0.36	[80]
	0,41		0.34	[71]
Nitrobenzene	0.49±0.06	—	0.23	Our data
Quinoline	0.32±0.05	—	—	" "
Benzaldehyde	0.37±0.07	—	—	" "
Aniline	0.39±0.05	—	0.25	" "

in [8, 22] that the quantity $\left(\rho\frac{\partial\varepsilon}{\partial\rho}\right)$ which occurs in the coefficient L in Eq. (1.12) is best calculated using the "low"* values of the scattering coefficient R or the dynamic value of this quantity should be used. Table 7 lists the results of calculations based on Eq. (1.12), in which the "low" values of the scattering coefficient R were used. We can see that this equation agrees better with the experimental results than does Eq. (1.11). The same conclusion is reported in [80].

The lack of the necessary data made it impossible to carry out calculations by means of Eq. (1.12) and, in some cases, even by means of Eq. (1.11); this is why the calculated values in Table 7 are incomplete.

CHAPTER IV

TRANSVERSE HYPERSONIC WAVES IN LOW-VISCOSITY LIQUIDS

§12. Narrow Wing of the Rayleigh Line and Fluctuations of the Anisotropy Due to Shear Strains

In the preceding chapters we considered the principal phenomena associated with the scattering of light by adiabatic fluctuations of the density. However, we must remember that fluc-

---

* The large number of the available values of the scattering coefficient R can be divided into two group known as the "high" and "low" values, which differ by up to ~40% [1].

tuations of the molecular orientation or fluctuations of the anisotropy also make a large contribution to the total flux of the scattered light. Fluctuations of the density are understood to mean fluctuations of the number of particles in a fluctuation volume, whereas fluctuations of the anisotropy imply some order in the orientation of the molecules within the fluctuation volume. We shall now consider the spectrum of light scattered by fluctuations of the anisotropy and restrict our treatment to a narrow range (< 1 cm^{-1}) of frequencies adjoining the exciting line.

The scattering of light by fluctuations of the anisotropy gives rise to the wing of the Rayleigh line. Landau and Placzek [4] were the first to show that this wing is due to the relaxation phenomena occurring in the liquid. The width of the wing corresponds approximately to the reciprocal of the relaxation time of the dipole moment [42].

The phenomenological theories of Leontovich [5] and Rytov [7] deal with the scattering of light by those fluctuations of the anisotropy which are due to fluctuations of strain. Consequently, these theories are applicable to that part of the spectrum which is known as the "shear" wing. On the other hand, the semiphenomenological theory of Starunov [8] describes the whole spectrum of the wing of the Rayleigh line, with the exception of those cases and those frequencies at which shear waves appear in the liquid (this applies to the internal parts of the Brillouin components of liquids with fairly high viscosities).

Fabelinskii [1, 6, 34] was the first to investigate the narrow part of the wing of the Rayleigh line adjoining directly the exciting line. An investigation of the temperature dependence of the half-width of the narrow part of the wing of the Rayleigh line and the depolarization of this part in the case of low-viscosity liquids (carbon disulfide, acetic acid, benzene, etc.) indicated that the narrow part could be described satisfactorily by the rotational diffusion mechanism. Therefore, the relaxation time τ should be given by Eq. (3.13). Fabelinskii [6], Starunov [8], and other workers [1] compared the experimental values of the relaxation time τ and the values calculated from Eq. (3.13) using the published data on the dimensions of molecules. It was found that the experimental and calculated values of τ were in good agreement [1]. Hence, it was concluded that the correct theoretical interpretation of the wing of the Rayleigh line of low-viscosity liquids was provided by Eq. (3.13).

An increase in the viscosity of the liquid should enhance the contribution of the shear wing of the Rayleigh line. The part of the wing close to the exciting line should become narrower and one would need light sources with narrow exciting lines and high-resolution spectrographs in studies of this part.

The use of a gas laser ($\lambda = 6328$ Å) provided basically new opportunities for simultaneous investigations of the wing and of the fine structure lines. The interference unit described in Chap. II was used to separate the scattered light into two components with mutually perpendicular polarizations. In this way, we investigated the narrow parts of the wing of the Rayleigh line of nitrobenzene, quinoline, benzaldehyde, and aniline. All these liquids were fairly viscous (> 1 cP). The molecules in these liquids were known to be strongly anisotropic. The depolarization of the Rayleigh-scattered light in nitrobenzene, quinoline, benzaldehyde, and aniline was about 50%. This enabled us to use the same photographs of the Rayleigh scattering spectra in studies of the fine structure components and of the wing of the Rayleigh line.

Figure 17 shows the spectrum of the Rayleigh scattering of light in nitrobenzene, obtained for light polarized along the z axis. Figure 18 shows a similar spectrum of the light scattered in quinoline. It is evident from Fig. 17 that the narrow part of the wing of the Rayleigh line is superimposed on the fine structure components (the z-polarized scattered light). We studied the influence of temperature on the narrow part of the wing of the Rayleigh line of benzene and the results of this study are plotted in Fig. 19. If this part of the wing is due to the rotational diffusion mechanism, the dependence of dν on the reduced temperature T/η should be linear, in

Fig. 17. Interferogram of the light scattered in nitrobenzene (exciting light polarized along the z axis): c is the central line; B are the Brillouin components; FSWR is the fine structure of the wing of the Rayleigh line.

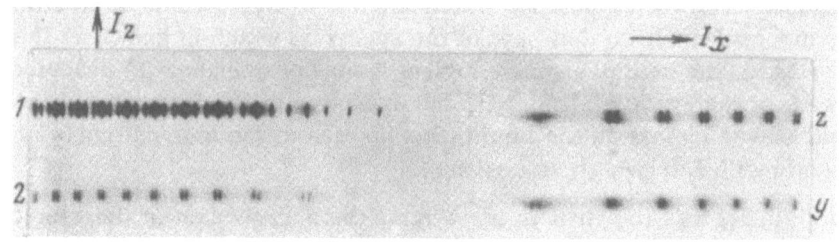

Fig. 18. Interferograms of the light scattered in quinoline obtained using exciting light with two polarizations: 1) along the z axis; 2) along the y axis.

accordance with Eq. (3.13). It is evident from Fig. 19 that the half-width $d\nu$ does indeed depend linearly on the ratio T/η. Consequently, our measurements support the conclusions of Fabelinskii [1, 6, 34] and Starunov [8] that the narrow part of the wing of the Rayleigh line, adjoining the exciting line, can be attributed to the rotational diffusion mechanism.

The line plotted in Fig. 19 can be used to determine the ratio

$$\frac{\Delta (d\nu)}{\Delta (T/\eta)} = \frac{1}{\pi^2 c \frac{4}{3} \frac{a^3}{k}} .$$

The dimensions of the nitrobenzene molecule can be deduced from the above ratio. The appropriate calculations indicate that $a = 3.2 \times 10^{-8}$ cm. This value is in agreement (within the limits of the experimental error) with the calculations based on other methods [32].

A detailed investigation of the narrow parts of the wing of the Rayleigh line of nitrobenzene, quinoline, and aniline revealed a new phenomenon, which is the fine structure in the wing of the Rayleigh line. This fine structure is a doublet (Figs. 17 and 18) polarized in a certain manner. When the incident light is polarized along the z axis, the doublet is observed in the x polarization component of the scattered light, but if the incident light is polarized along the y axis, this doublet is observed in the z polarization component. The dip in the intensity between the two parts of the doublet does not decrease to zero and it corresponds to the frequency of the undisplaced component polarized along the z axis (in the case when the incident light is polarized along the z axis). This phenomenon can be explained if we bear in mind that the light scattered by fluctuations of the anisotropy associated with fluctuations of strain is modulated by a transverse acoustic wave ($\sim 10^9$ Hz).

The formulas (3.5) deduced phenomenologically by Leontovich show that when the scatter-

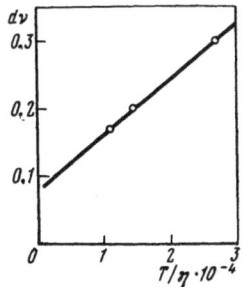

Fig. 19. Dependence of the half-width of the wing of the Rayleigh line of nitrobenzene dν on the ratio D/η.

ing is excited by light polarized along the z axis, the x polarization component of the scattered light should have a doublet at $\omega = \pm\Omega_T$ due to the influence of shear strains on the anisotropy of the liquid in question. The formulas (3.5) show that the intensity at the frequency of the exciting line should have a minimum which does not fall to zero because the component $S_A(\omega)$, which has a strong influence on the general distribution of the intensity, has a maximum at this frequency. It is evident from Eqs. (3.5) and (3.9) that the separation between the frequencies of the components of the doublet is

$$\Omega_T = 2n\omega_0 \frac{v_T}{c} \sin\frac{\theta}{2}, \tag{12.1}$$

which includes the velocity of the transverse hypersound

$$v_T = \sqrt{\mu/\rho}, \tag{12.2}$$

where μ is the shear modulus. However, because of the influence of the term $S_A(\omega)$, the observed frequency shift of the components of the doublet $\Omega_{\max}$ should be less than the frequency of the transverse hypersonic waves Ω_T.

The doublet can be observed under optimal conditions under which Ω_T and τ have values such that the modulation by the transverse acoustic wave is not masked by the influence of the part of the wing with its peak at $\omega = 0$ and by the transfer function of the spectrograph.

A comparison of Eqs. (3.5) and (3.6) shows that when the polarization of the exciting light is changed from the z to the y axis, the doublet observed in the component of the scattered light with the x polarization should switch to the z component of the scattered light, i.e.,

$$I_{zx}(\omega) = I_{yz}(\omega).$$

Figure 20 shows the experimentally determined intensity of the x polarization component of the light scattered in quinoline. The theoretical (continuous) curve is plotted using the measured values of Ω_T and τ.

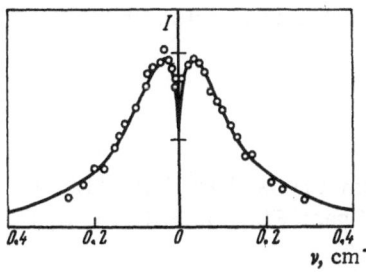

Fig. 20. Distribution of the intensity in the wing of the Rayleigh line of quinoline. The continuous curve represents calculations based on the Leontovich theory [5]: the open circles represent the experimental values.

TABLE 8

Liquid	t, °C	$\eta \cdot 10^2$, P	$2\Delta\nu_{max}$, cm^{-1}	$f \cdot 10^{-9}$, Hz	v_T, m/sec	$\mu \cdot 10^{-9}$, dyn/cm^2
Quinoline	20	3.60	0.078	1.20	320	1.2
Nitrobenzene	20	2.03	0.043	0.64	180	0.4
Aniline	20	4.40	0.130	1.95	550	3.1

The phenomenological theory of the scattering of light describes quantitatively the polarization of the fine structure in the wing of the Rayleigh line.

Investigations of the angular dependence of the shift of the components of the fine structure of the Rayleigh line would provide important evidence for or against the proposed interpretation of this fine structure. Our study showed that the approximate separation between the components did indeed vary proportionally to $\sin \theta/2$; however, the precision of these measurements, particularly for scattering angles $\theta < 90°$, was not high because the separation between the components was small and their width was of the same order as the separation. Therefore, an allowance for the influence of the width of the components on their position would be needed in high-precision measurements.

If we measure $\Delta\nu_{max} = \Omega_T/2\pi c$, we can apply Eq. (12.1) to find the velocity of transverse sound v_T at the frequency Ω_T. It follows from Eq. (12.2) that the velocity v_T can be used to find the shear modulus μ.

Table 8 lists the results of the first measurements of the velocity of the transverse sound and of the shear modulus, deduced from the separation between the components of the fine structure in the wing of the Rayleigh line ($\theta = 90°$).

In this first investigation no correction was made for the influence of the term $S_A(\omega)$ and for the width of the components on the frequency Ω_{max}. Estimates showed that the anisotropy relaxation time of nitrobenzene and quinoline, given by Eq. (3.12) and the relaxation time of the shear viscosity, given by Eq. (3.2), were of the same order of magnitude and had values in the range $\tau \sim (2-5) \times 10^{-11}$ sec. Consequently, in our case, $\Omega_T\tau \approx 1$ and the correction to the values of v_T listed in Table 8 could reach $\sim 10-15\%$. These estimates also indicated that the transverse waves could travel in a liquid but would be damped quite strongly. Since $\Omega_T\tau \approx 1$, the absorption coefficient per wavelength would be $\alpha\Lambda \approx \pi$ [81]. Consequently, it would be difficult or impossible to observe the transverse components in the scattered light because their half-widths would be of the same order as the separation between them. The theories of Leontovich [5] and Rytov [7] predict that the transverse hypersonic waves should influence fluctuations of the anisotropy in the scattering of light. However, a more definite conclusion about the influence of the damping of the transverse hypersonic waves on the distribution of intensity in the shear wing cannot be drawn because the theories of Leontovich [5] and Rytov [7] ignore the influence of the damping of the transverse waves on the intensity distribution.

The author is deeply grateful to I. L. Fabelinskii and V. S. Starunov for suggesting the subject, supervising the work, and for their constant advice. He is also grateful to V. P. Zaitsev, M. A. Vysotskaya, and S. V. Krivokhizh for their great help in this investigation.

LITERATURE CITED

1. I. L. Fabelinskii, Molecular Scattering of Light, Plenum Press, New York (1968).
2. L. I. Mandel'shtam, Zh. Russ. Fiz.-Khim. Obshchst., Chast'Fiz., 58:381 (1926); Complete Works [in Russian], Vol. 1, Izd. AN SSSR, Moscow (1957).

3. L. Brillouin, Ann. Phys. (Paris), 17:88 (1922).
4. L. Landau and G. Placzek, Phys. Z. Sowjetunion, 5:172 (1934).
5. M. Leontovich, J. Phys. USSR, 4:499 (1941).
6. I. L. Fabelinskii, Usp. Fiz. Nauk, 63:355 (1957); Tr. Fiz. Inst. Akad. Nauk SSSR, 9:181 (1958).
7. S. M. Rytov, Zh. Eksp. Teor. Fiz., 33:514, 671 (1957).
8. V. S. Starunov, Dokl. Akad. Nauk SSSR, 153:1055 (1963); Opt. Spektrosk., 18:300 (1965); Dissertation [in Russian], Physics Institute, Academy of Sciences of the USSR, Moscow (1965).
9. G. S. Landsberg, Selected Papers [in Russian], Izd. AN SSSR, Moscow (1958).
10. E. Gross, Z. Phys., 63:685 (1930).
11. E. Gross, Nature, 126:201, 400, 603 (1930).
12. L. I. Mandel'shtam and M. S. Leontovich, Zh. Eksp. Teor. Fiz., 7:438 (1937).
13. I. L. Fabelinskii and O. A. Shustin, Dokl. Akad. Nauk SSSR, 92:285 (1953).
14. V. A. Molchanov and I. L. Fabelinskii, Dokl. Akad. Nauk SSSR, 105:248 (1955).
15. M. S. Pesin and I. L. Fabelinskii, Dokl. Akad. Nauk SSSR, 122:575 (1958).
16. I. L. Fabelinskii, Usp. Fiz. Nauk, 77:649 (1962).
17. M. S. Pesin and I. L. Fabelinskii, Dokl. Akad. Nauk SSSR, 129:299 (1959).
18. D. H. Rank and A. E. Douglas, J. Opt. Soc. Amer., 38:966 (1948).
19. M. I. Shakhparonov and M. S. Tunin, Applications of Ultrasound in Investigations of Matter [in Russian], Moscow Regional Pedagogical Institute (1961).
20. M. S. Tunin, Dissertation [in Russian], Moscow Regional Pedagogical Institute (1962).
21. D. I. Mash, V. S. Starunov, and I. L. Fabelinskii, Zh. Eksp. Teor. Fiz., 47:783 (1964).
22. D. I. Mash, V. S. Starunov, E. V. Tiganov, and I. L. Fabelinskii, Zh. Eksp. Teor. Fiz., 49:1764 (1965).
23. H. Lamb, Hydrodynamics, 6th ed., Cambridge University Press (1932).
24. M. Leontovich (Leontowitsch), Z. Phys., 72:247 (1931).
25. M. A. Leontovich, Izv. Akad. Nauk SSSR, Ser. Fiz., No. 5, p. 633 (1936).
26. L. D. Landau and E. M. Lifshitz, Electrodynamics of Continuous Media, Pergamon Press, Oxford (1960).
27. J. B. Lastovka and G. B. Benedek, Phys. Rev. Lett., 77:1039 (1966).
28. V. V. Vladimirskii, Zh. Eksp. Teor. Fiz., 9:1226 (1939).
29. V. L. Ginzburg (Ginsburg), Dokl. Akad. Nauk SSSR, 42:168 (1944); Izv. Akad. Nauk SSSR, Ser. Fiz., 9:174 (1945).
30. E. F. Gross, Zh. Eksp. Teor. Fiz., 16:129 (1946).
31. L. D. Landau and E. M. Lifshitz, Fluid Mechanics, Pergamon Press, London (1959).
32. L. Bergmann, Ultrasonics, Bell, London (1938) [transl. from Ultraschall, VDI-Verlag, Berlin (1937)].
33. M. A. Isakovich, Dokl. Akad. Nauk SSSR, 23:783 (1939).
34. V. K. Ablekov and I. L. Fabelinskii, Dokl. Akad. Nauk SSSR, 125:297 (1959).
35. G. B. Benedek, J. B. Lastovka, K. Fritsch, and T. J. Greytak, J. Opt. Soc. Amer., 54:1284 (1964).
36. J. E. Piercy and G. R. Hanes, J. Chem. Phys., 43:3400 (1965).
37. G. (B.) Benedek and T. (J.) Greytak, Proc. IEEE, 53:1623 (1965).
38. T. J. Greytak and G. B. Benedek, Phys. Rev. Lett., 17:179 (1966).
39. J. B. Lastovka and G. B. Benedek, Phys. Rev. Lett., 17:1039 (1966).
40. R. Y. Chiao and P. A. Fleury, in: Physics of Quantum Electronics (Proc. Intern. Conf., San Juan, Puerto Rico, 1965), publ. by McGraw-Hill, New York (1966), p. 241; P. A. Fleury and R. Y. Chiao, J. Acoust. Soc. Amer., 39:751 (1966).
41. E. Garmire and C. H. Townes, Appl. Phys. Lett., 5:84 (1964).
42. D. I. Mash, V. V. Morozov, V. S. Starunov, E. V. Tiganov, and I. L. Fabelinskii, ZhETF Pis. Red., 2:246 (1965).

43.   A. A. Chaban, ZhETF Pis. Red., 3:73 (1966).
44.   R. G. Brewer, Appl. Phys. Lett., 9:51 (1966).
45.   T. S. Velichkina, O. A. Shustin, and I. A. Yakovlev, ZhETF Pis. Red., 2:189 (1965).
46.   E. V. Tiganov, Prib. Tekh. Eksp., No. 4, p. 240 (1967).
47.   V. P. Zaitsev, Prib. Tekh. Eksp., No. 3, p. 214 (1966).
48.   S. Tolansky, High Resolution Spectroscopy, Methuen, London (1947).
49.   V. S. Starunov, E. V. Tiganov, and I. L. Fabelinskii, ZhETF Pis. Red., 5:317 (1967).
50.   D. H. Rank, E. R. Shull, and D. W. E. Axford, Nature, 164:67 (1949).
51.   N. A. Clark, C. E. Moeller, J. A. Bucaro, and E. F. Carome, J. Chem. Phys., 44:2528 (1966).
52.   Technical Encyclopedia [in Russian], GRTÉiS, Moscow (1946).
53.   S. V. Krivokhizha and I. L. Fabelinskii, Zh. Eksp. Teor. Fiz., 50:3 (1966).
54.   V. S. Starunov, E. V. Tiganov, and I. L. Fabelinskii, ZhETF Pis. Red., 4:262 (1966).
55.   R. Y. Chiao and B. P. Stoicheff, J. Opt. Soc. Amer., 54:1286 (1964).
56.   P. A. Bazhulin, Tr. Fiz. Inst. Akad. Nauk SSSR, 5:261 (1950).
57.   E. L. Heasell and J. Lamb, Proc. Phys. Soc., London, B69:869 (1956).
58.   J. L. Hunter, E. F. Carome, H. D. Dardy, and J. A. Bucaro, J. Acoust. Soc. Amer., 40:313 (1966).
59.   N. B. Lezhnev, Dissertation [in Russian], Physicotechnical Institute, Academy of Sciences of the Tadzhik SSR, Ashkhabad (1963).
60.   M. S. Pesin, Tr. Fiz. Inst. Akad. Nauk SSSR, 30:158 (1964) [Physical Optics, Consultants Bureau, New York (1966), pp. 125-175].
61.   D. Sette, A. Busala, and J. C. Hubbard, J. Chem. Phys., 23:787 (1955).
62.   N. J. Meyer, J. Chem. Phys., 33:487 (1960).
63.   J. L. Hunter and H. D. Dardy, J. Chem. Phys., 44:3637 (1966).
64.   G. R. Hanes, R. Turner, and J. E. Piercy, J. Acoust. Soc. Amer., 38:1057 (1965).
65.   J. L. Hunter and H. D. Dardy, J. Chem. Phys., 42:2961 (1965).
66.   L. V. Lanshina, Yu. G. Shoroshev, and M. I. Shakhparonov, Dokl. Akad. Nauk SSSR, 173:70 (1967).
67.   B. V. Raghavendra Rao, Nature, 139:885 (1937); Proc. Indian Acad. Sci., A3:163 (1936).
68.   B. V. Raghavendra Rao and D. S. Subba Ramaiya, Phys. Rev., 60:615 (1941).
69.   V. L. Ginzburg, Dokl. Akad. Nauk SSSR, 36:8 (1942).
70.   S. L. Shapiro, M. McClintock, D. A. Jennings, and R. L. Barger, IEEE J. Quantum Electron., QE-2:89 (1966).
71.   D. H. Rank, E. M. Kiess, U. Fink, and T. A. Wiggins, J. Opt. Soc. Amer., 55:925 (1965).
72.   J. H. Andreae, Proc. Phys. Soc., London, B70:71 (1957).
73.   E. S. Stewart and J. L. Stewart, Phys. Rev. Lett., 13:437 (1964).
74.   E. V. Tiganov, ZhETF Pis. Red., 4:385 (1966).
75.   E. L. Gordon and M. G. Cohen, Phys. Rev., 153:201 (1967).
76.   L. Hall, Phys. Rev., 73:775 (1948).
77.   A. Eucken, Z. Elektrochem., 52:255 (1948).
78.   K. F. Herzfeld and T. A. Litovitz, Absorption and Dispersion of Ultrasonic Waves, Academic Press, New York (1959).
79.   C. L. O'Connor and J. P. Schlupf, J. Acoust. Soc. Amer., 40:663 (1966).
80.   H. Z. Cummins and R. W. Gammon, J. Chem. Phys., 44:2785 (1966).
81.   J. Frenkel, Kinetic Theory of Liquids, Oxford University Press (1946).

STIMULATED MOLECULAR SCATTERING
OF LIGHT IN GASES *

V. V. Morozov

Stimulated Brillouin scattering in compressed gases was investigated using a ruby laser as the excitation source. The time evolution of the Brillouin scattering in compressed nitrogen was studied. Different detection systems were used in a spectroscopic investigation of the scattering as a function of the intensity of the exciting radiation. Stimulated Brillouin and temperature (entropy) scattering processes were observed in low-pressure gaseous hydrogen. Laser radiation was found to undergo anti-Stokes broadening as a result of interaction in various gases. The results obtained were compared with the predictions of currently available theories.

INTRODUCTION

The scattering of light by random fluctuations of the density, orientation, and concentration of molecules is known as the molecular scattering [1], in contrast to the Raman scattering and the scattering by colloidal and other suspended particles.

Experimental and theoretical investigations of the molecular scattering spectra of gases have yielded extensive information on the kinetics of thermal fluctuations and on acoustic and other properties of gases at frequencies which are not attainable in other methods.

The difference between the exciting-radiation and the scattered-light spectra is due to the modulation of the scattered light because of variation in the fluctuations of the density with time, which give rise to fluctuations in the optical permittivity.

Fluctuations of the density in a gas can be regarded as consisting of fluctuations of the pressure (adiabatic or isentropic fluctuations of the density) and fluctuations of the entropy (isobaric fluctuations of the density).

A fairly comprehensive phenomenological theory of the scattering of light by fluctuations of the density has been developed [1-15]. In particular, this theory gives a quantitative expression for the distribution of the intensity in the spectrum of the scattered light. Modulation of the scattered light by variation of the pressure fluctuations with time or, in other words, modulation of the scattered light wave by changes in the refractive index due to the propagation of an

---

* Based on a thesis submitted for the degree of Candidate of Physicomathematical Sciences, defended on October 21, 1968, at the P. N. Lebedev Physics Institute, Academy of Sciences of the USSR. Scientific supervisors: I. L. Fabelinskii and D. I. Mash.

elastic wave gives rise to frequency-shifted Brillouin components in the scattered-light spectrum.*   ๋

Modulation of the scattered light by isobaric fluctuations of the density, which vary with time, is responsible for the undisplaced (central) component in the scattered-light spectrum.

The theoretical formulas which described the fine-structure spectrum can be used to deduce the velocity of hypersound of $\sim 10^9$ Hz frequency (from the relative positions of the Brillouin components) and the coefficient of absorption of hypersound (from the widths of these components).

However, in order to obtain such information from the fine structure of scattered light, the exciting radiation must be of high intensity and very narrow wavelength. The absence of light sources which would satisfy these requirements was a very serious difficulty in the early experimental investigations of gases.

For these reasons the fine structure of the Rayleigh line of gases was not discovered and investigated spectroscopically until 1965. The only experimental study [16] carried out before the appearance of lasers gave results which were in basic disagreement with the theory [7, 17-19]. Lasers made it possible to study experimentally the stimulated scattering of light in gases and to investigate the fine structure of the Rayleigh line.

Mash, Starunov, Fabelinskii, and the present author [20] reported the discovery of a fine structure in the stimulated Brillouin scattering in compressed nitrogen, hydrogen, and oxygen. Independently of us, Rank et al. [21] and Hagenlocker and Rado [22], working in two separate laboratories in the USA, discovered the same phenomenon in nitrogen, methane, carbon dioxide, and argon.

The discovery of the stimulated Brillouin scattering in compressed gases showed clearly that the earlier failures were not due to some misunderstanding of the nature of the phenomenon [16] but due to insufficiently fine experimental techniques. This was confirmed by the more recent discovery and investigations of the fine structure of the Rayleigh line in the thermal scattering [23] excited by the $\lambda = 6328$ Å gas-laser line.

In the thermal scattering of light the weak field of the exciting radiation has a negligible influence on the medium in question, but the situation is quite different in the case of propagation of a giant pulse produced by a ruby laser. In the latter case, the electric field of the exciting radiation is so strong that, together with the field of the scattered wave, it interacts with the medium and gives rise to a striction force which results in stimulated scattering of the Stokes component of the Brillouin radiation [1]. In this effect the intensity of the Stokes component of the fine structure of the scattering line increases nonlinearly with increasing size of the scattering region provided the intensity of the exciting radiation is higher than a certain threshold value.

A new nonlinear phenomenon in the form of stimulated temperature (entropy) scattering of light in gaseous hydrogen was discovered by Fabelinskii, Mash, Starunov, and the present author [24]. The nonlinearity of the interaction between the exciting laser wave and the scattered temperature waves, resulting from fluctuations in the entropy, is due to the electrocaloric effect.

The spectrum of the stimulated temperature scattering differs considerably from the thermal scattering spectrum. Instead of a monotonic fall on both sides of the exciting line in the thermal scattering (this fall is due to fluctuations of the entropy), the stimulated scattering

---

* Brillouin scattering is usually referred to as the Mandel'shtam−Brillouin scattering in the Soviet literature.

spectrum contains only the Stokes component and a very important peak at the half-width of the thermal scattering line.

Current studies of the thermal and stimulated molecular scattering of light in different states of matter are characterized by considerable improvements in the laser technology, experimental methods, and theoretical investigations of the thermal and nonlinear molecular scattering processes. As usual, many questions have remained unanswered. The present paper reports the results of experimental investigations and theoretical analyses of some of the unsolved problems in the stimulated molecular scattering of light in gases.

The first task in the experimental investigations is the provision of a source of sufficiently high power emitting in a very narrow spectral range. The present author used a Q-switched ruby laser in which ordinary crystals emitted single pulses up to 300 MW power in a spectral range of ≤ 0.01 cm^{-1}.

This source of light was used in the discovery of stimulated Brillouin and temperature scattering processes in gases. Moreover, in an earlier study the same source was employed in the discovery of a stimulated wing of the Rayleigh line and in various other experiments. The same type of laser was used successfully in other investigations.

The first determination of the velocity of hypersound from the Brillouin components of the stimulated scattering indicated that at $\sim 10^9$ Hz this velocity was close to the isothermal value, in spite of the fact that the measurements were carried out under conditions such that the frequency at which the adiabatic propagation changed to the isothermal process ($\sim 10^{11}$ Hz) was considerably higher than the frequency at which significant scattering was observed. Moreover, spectrograms obtained in this study included up to four Stokes components and one anti-Stokes component of the stimulated Brillouin scattering line.

A special unit with a time resolution of $\sim 1.5 \times 10^{-9}$ sec was developed. This unit was used to study the kinetics of the appearance of the stimulated Brillouin scattering lines and to measure simultaneously the integrated intensities of the light scattered by and transmitted through the interaction region. This and the subsequent study of the stimulated Brillouin scattering in compressed nitrogen under integrating observation conditions made it possible to determine the mechanisms responsible for the appearance of a large number of Brillouin components in the stimulated scattering and to find the dependence of the shift of these components on the intensity of the exciting radiation in the interaction region (Chap. III).

The method developed for additional amplification of the components of the radiation scattered at $\theta = 180°$ was based on the use of a superregenerative amplifier or, which is equivalent, a trigger amplifier utilizing the laser resonator modes (Chap. II).

This method made it possible to study the appearance of the Brillouin scattering components which could not have been observed by other methods in low-pressure gases. The dependence of the shift of these components on the pressure in gaseous hydrogen was determined. The results of this investigation were in agreement with the conclusions of the gas-kinetic theory.

A study was also made of a new nonlinear effect discovered by the author. This effect was the stimulated temperature (entropy) scattering of light in hydrogen. This discovery was made possible by the use of additional trigger amplification of the stimulated scattering of light (Chap. IV).

A study was also made of the frequency broadening of laser pulses in gases and the possible causes of such broadening are considered.

CHAPTER I

THEORETICAL AND EXPERIMENTAL INVESTIGATIONS
OF THE SPECTRA OF MOLECULAR SCATTERING OF
LIGHT IN GASES

§1. General Aspects of the Theory of Molecular Scattering of Light in Gases (Continuous Medium Approximation)

The application of hydrodynamic equations to calculations concerned with the spectra of the molecular scattering of light in isotropic media is justified if the mean free path of the molecules ($\bar{l}$) is much shorter than the wavelength (Λ) of the Fourier components of the fluctuations responsible for the scattering of light.

If the condition $\bar{l} \ll \Lambda$ is satisfied, the gaseous medium can be regarded as continuous and it can be represented by the optical value of the permittivity ε.

The thermal motion of molecules in a gas gives rise to fluctuations of the density and orientation of anisotropic molecules and these, in their turn, generate fluctuations of the optical permittivity.

Fluctuations of the pressure, temperature (or entropy), and orientation of anisotropic molecules or fluctuations of the concentration continuously appear and disappear. The nature of this process of appearance and disappearance of the fluctuations varies with the nature of the fluctuations.

The first theoretical investigation of the spectrum of the modulation of scattered light was carried out by Mandel'shtam [2] in 1918, although he published his results much later (1926). Meanwhile, Brillouin [3] derived independently and published some of the results given in [2].

These investigations of Mandel'shtam and Brillouin demonstrated that the spectrum of monochromatic light of frequency ν scattered by adiabatic fluctuations of the density should contain not only the frequency ν but also shifted frequencies (fine structure of the Rayleigh line).

Mandel'shtam, Landsberg [4], and Gross [5] discovered experimentally the predicted effects in a crystal of quartz and Gross also found the fine structure of the Rayleigh line in a liquid [6].

The question of whether the fine structure of the Rayleigh line can be observed in gases above atmospheric pressure ($\bar{l} \ll \Lambda$) remained unsolved up to 1965 [7]. We shall consider this question in more detail later.

Spectral Composition of Light Scattered by Pressure Fluctuations

If a medium is illuminated with a parallel beam of light of definite wave vector $\mathbf{K}$ and the scattered light is observed in the direction of a wave vector $\mathbf{K}'$, a maximum of the intensity of the scattered (or diffracted) light is observed only if the wave vector of an elastic wave $\mathbf{q}$ satisfies the Bragg condition in conjunction with the wave vectors $\mathbf{K}$ and $\mathbf{K}'$:

$$\mathbf{K}' - \mathbf{K} - \mathbf{q} = 0. \qquad (\text{I}.1)$$

Thus, if we specify the directions of **K** and of **K'** (or the angle of observation θ), we select from an extensive set just two waves with identical frequencies Ω and with wave vectors **q** of the same magnitude but opposite in direction. The absolute value of $|\mathbf{q}|$ is then given by the relationship

$$|\mathbf{q}| = 4\pi n/\lambda \cdot \sin \theta/2 = 2\pi/\Lambda, \tag{I.2}$$

where Λ is the wavelength of the elastic wave.

An optical inhomogeneity which varies with time must modulate the scattered light. We shall start by considering the modulation resulting from adiabatic fluctuations of the density or pressure.

If the function which modulates the scattered light represents a monochromatic Debye acoustic wave of wavelength Λ, the scattered light has two Brillouin components [1], which are separated from the exciting line of frequency ω_0 by

$$\Omega_0 = vq = 4\pi nv/\lambda \sin \theta/2 = 2\omega_0 nv/c \sin \theta/2; \tag{I.3}$$

here, the relative frequency shift of the Brillouin components or satellites is

$$\pm \frac{\Delta\omega}{\omega_0} = \pm \frac{\Omega_0}{\omega_0} = \pm \frac{\Delta v}{v} = 2nv/c \sin \theta/2. \tag{I.4}$$

The formula (I.4) was derived first by Mandel'shtam [2] and Brillouin [3]. It follows from this formula that the relative shift of the frequency depends only on the ratio v/c, the angle of observation θ, and the refractive index n.

So far, we have assumed that the elastic wave is not absorbed and, therefore, the Brillouin components are infinitely narrow (if they are excited by a monochromatic line of frequency ω_0). In fact, the elastic wave is attenuated to a smaller or greater extent in any material medium and, therefore, the fine-structure components always have finite widths.

The influence of the absorption of the elastic wave on the width of the Brillouin components was first considered by Leontovich [8, 9] and later by other workers [10-13]. In this case, a Debye wave can be represented in the form

$$\Phi(t) = \Phi_0 e^{-\delta_1 t} \cos[\Omega t - \mathbf{q}\mathbf{r}],$$

where $\delta_1 = \Gamma q^2/2$; $\Gamma = 1/\rho\{4/3\eta + \eta'\}$; η and η' are the shear and bulk viscosities of the medium.

Appropriate calculations [1] show that the spectral distribution of the intensity of light scattered by adiabatic fluctuations of the density is

$$I(\omega_0 \pm \Omega) = \Omega_0^2 \frac{2\delta_1}{\pi} \Big/ (\Omega_0^2 - \Omega^2)^2 + 4\delta_1^2\Omega^2, \tag{I.5}$$

where Ω is the angular frequency measured from the unshifted line. The intensity maximum of a Brillouin component can be found from Eq. (I.5):

$$\Omega_{max} = (\Omega_0^2 - 2\delta_1^2)^{1/2}. \tag{I.6}$$

If $\delta_1/\Omega \ll 1$, the half-width of such a component is

$$\delta\Omega = 2\delta_1 \left(1 + \frac{1}{2}\frac{\delta_1^2}{\Omega^2}\right) \approx 2\delta_1 = 2\alpha v, \qquad (I.7)$$

where α is the amplitude absorption coefficient of the elastic thermal wave.

If we combine Eqs. (I.3) and (I.7), we obtain the condition for resolution of the components:

$$\alpha\Lambda \ll \pi. \qquad (I.8)$$

If the absorption coefficient of ultrasound is extrapolated quadratically to the frequencies of the Brillouin components ($f = 10^9$-10^{10} Hz), it is found that the condition $\alpha\Lambda \ll \pi$ is not satisfied by benzene and other liquids and by compressed gases. However, experimental investigations of liquids and later studies of compressed gases indicated that the fine structure of the Rayleigh line could be observed quite clearly.

In the case of gases, this contradiction is resolved by an allowance for the Kneser effect. In this effect the energy is redistributed between the external (translational and rotational) and the internal (vibrational) degrees of freedom of molecules. The theory of the influence of this effect on the propagation of sound in gases was formulated clearly by Kneser [14]. Later Mandel'shtam and Leontovich [15] developed a very general relaxation theory of the propagation of acoustic waves in condensed media.

Spectral Composition of Light Scattered by Entropy (Temperature) Fluctuations

The scattering of light by isobaric fluctuations of the density can be analyzed in exactly the same way as the scattering by adiabatic fluctuations of the density. However, in the present case, the function which modulates the scattered light is quite different:

$$\Phi(t) = \Phi_0 \exp\left[-(\delta_2 t + i\mathbf{q}\mathbf{r})\right], \qquad (I.9)$$

where

$$\delta_2 = \chi q^2 \qquad (I.10)$$

and χ is the thermal diffusivity.

It follows from Eq. (I.9) that $\Phi(t)$ decays exponentially with time and that, in contrast to elastic waves, optical inhomogeneities occupy fixed positions in space.

The half-width of the central line of the fine structure is now given by

$$\delta\omega_c = 2\chi\mathbf{q}^2. \qquad (I.11)$$

§2. Fine Structure of the Rayleigh Scattering of Light in Gases ($\bar{l} \ll \Lambda$)

Only one experimental investigation of the fine structure of the Rayleigh line was carried out before 1965: this was the investigation carried out by Venkateswaran [16] in hydrogen (H_2) at 100 atm, in nitrogen (N_2) and oxygen (O_2) at 80 atm, and in carbon dioxide (CO_2) at 50 atm.

The scattered light was observed at an angle of $\theta = 180°$. The gap between the Fabry–Perot interferometer mirrors was $t_1 = 7.5$ mm. A discrete fine structure of the Rayleigh line was not observed in any of the four gases under investigation. Only some broadening of the scattered-light line was observed for H_2; in the other three cases, the spectrum of the scattered light was identical with the spectrum of the exciting radiation. These results were reviewed critically in [7].

According to Venkateswaran [16], the spectrum of the light scattered in all four gases should not have any discrete fine structure but should exhibit broadening of the scattering line as a result of the Doppler effect associated with the thermal motion of the gas molecules.

The suggestion that line broadening in a gas kept at pressures of 50-100 atm can be due to the Doppler effect associated with single molecules moving at thermal velocities is probably due to a misunderstanding.

In fact, if a small (compared with the wavelength of light) region can be isolated in a medium containing molecules and the number of molecules in this region is proportional to its volume, we must consider the combined effect of scattering centers and represent the thermal motion of the molecules by Debye elastic waves. The frequency shift in the scattered light is then governed by the reflection from an elastic wave which acts as a mirror moving at a velocity ±v. A maximum of the scattered-light intensity should be observed along that direction which satisfies the Bragg condition and the frequency shift is due to the Doppler effect which is not associated with molecules moving at thermal velocities and distributed in accordance with the Maxwellian law but due to the reflection from elastic Debye waves. We have already shown that this effect leads to a discrete fine structure of the scattering line.

Landsberg [17] published a detailed critical analysis of the investigations aimed to detect the influence of the Doppler shift of the frequency of light as a result of the reflection from a metallic mirror. These investigations were carried out on the assumption that a frequency shift resulted from the thermal motion of atoms in the mirror material. Landsberg [17], Lennuier [18], and Cabannes [18] showed that, in this case, the replacement of the collective effect of atoms and molecules by the effect of isolated particles would lead to erroneous results.

A theory of the fine structure of the scattering line of a gas should be completely identical with the theory of the same phenomenon in a liquid as long as the mean free path $\bar{l}$ of the gas molecules is much shorter than the wavelength of a thermal elastic wave Λ which governs the scattering of light at a given angle. If hypersound of wavelength Λ is only slightly attenuated in a gas, the spectrum of light scattered by this gas will exhibit a clearly resolved fine structure. However, if the wavelength of hypersound is equal to or less than the mean free path of the gas molecules, the fine structure disappears and it is no longer possible to carry out a hydrodynamic analysis of the absorption of sound.

The condition specified in Eq. (I.8) means that the absorption of an elastic wave must be slight over a path equal to its wavelength. It is quite clear that the fine structure can be investigated experimentally only if the condition (I.8) is satisfied. This applies to the integrating interferometric investigation method [1].

The propagation of sound in gases at pressures employed by Venkateswaran [16] is adiabatic. The frequency of transition from the adiabatic to the isothermal propagation of sound is [1]

$$\Omega_{is} = v^2/\chi, \tag{I.12}$$

$$\chi = \varkappa/C_p\rho, \tag{I.13}$$

where $\varkappa$ is the thermal conductivity.

The frequency of the scattered light is given by Eq. (I.3):

$$\Omega_s = 2\pi v/\Lambda = 2\pi n v/\lambda \sin\theta/2. \tag{I.14}$$

We can easily see that, in the case of gases at pressures used in [16], the relevant ratio is

$$\Omega_{is}/\Omega_s = \lambda v/4\pi n\chi \sin\theta/2 \gg 1, \tag{I.15}$$

and, consequently, the propagation of hypersound is an adiabatic process. Therefore, we should substitute in Eq. (I.3) the Laplace velocity of sound in a gas, i.e.,

$$v = \sqrt{\gamma P/\rho}. \tag{I.16}$$

In the case of hypersound, we must allow for the dispersion of its velocity [15].

The hydrodynamic theory of the absorption of elastic waves in a viscous medium with a thermal conductivity $\varkappa$ yields the expression [1]

$$\alpha = q^2/2v\rho \{{}^4\!/_3\eta + \eta' + \varkappa/C_p(\gamma - 1)\}. \tag{I.17}$$

The experimentally obtained dependences of the absorption of sound on the ratio of the frequency of sound to the pressure (f/p) in gaseous carbon dioxide (CO_2) and hydrogen (H_2) [1] indicate that the bulk viscosity ceases to make any significant contribution to the absorption for ratios f/p = 1 MHz/atm and the whole absorption is due to the shear viscosity η. If we use Eqs. (I.17) and (I.16), substituting $\eta' = 0$ and employing a well-known gas-kinetic relationship, we obtain the following expression for the absorption of sound:

$$\alpha\Lambda = A\bar{l}/\Lambda, \tag{I.18}$$

where A = 25 for a diatomic gas and A = 24 for a triatomic gas.

It follows from Eq. (I.18) and from the numerical values of A that $\alpha\Lambda > \pi$ at atmospheric pressure ($\bar{l} \sim 10^{-5}$ cm) and, consequently, at this pressure the fine structure cannot be observed. However, when the pressure is increased to ~ 20 atm, we find that $\alpha\Lambda \ll \pi$ and $\alpha\Lambda \leq 0.1$ under the experimental conditions employed in [16], i.e., a well-resolved fine structure should be observed in the spectrum of the scattered light.

We may assume that Venkateswaran [16] failed to observe this structure because of the considerable experimental difficulties. In particular, the extremely low intensity of light made it necessary to use very long exposures. Venkateswaran used exposures of 15 days [16], taking measures to ensure a constant temperature in the room where the interferometer was located. However, no indication is given in [16] of any attempt to keep the pressure constant. If during the 14-day exposure the pressure between the interferometer plates (etalon thickness 7.5 mm and reflection coefficient 85%) changed by 2 mm Hg, such a change would have destroyed the discrete components in the spectrum of light scattered by gaseous hydrogen.

The classical phenomenological theory of the fine structure in a gas, which is outlined above, was known to be basically sound. On the other hand, the only experimental data obtained before the appearance of lasers [16] were in basic disagreement with the theory. Therefore, the experimental discovery of the fine structure in the Rayleigh line of the light scattered by gases had to await the advent of the lasers. In principle, this discovery could have been made before the development of the laser but it would have been very difficult.

The appearance of lasers has made it possible to study experimentally the scattering of light in gases and has stimulated the interest of investigators in the fine structure of the Rayleigh line.

In about a year from the discovery of the stimulated Brillouin scattering in compressed gases [20-22], the fine structure of the Rayleigh line was observed and studied in the thermal scattering spectra.

A low-power helium—neon gas laser made it possible to observe the fine structure of the Rayleigh line because the laser line $\lambda = 6328 \, \text{Å}$ was extremely narrow.

Eastman, Wiggins, and Rank [23] observed the fine structure of the Rayleigh line in compressed N_2 and CO_2 with the aid of a helium—neon laser emitting 250 mW. They used Eq. (I.8) in a high-precision determination of the velocity of hypersound, which was deduced from the positions of the shifted Brillouin components. The value of this velocity was in agreement with the theoretical predictions.

§3. Molecular Scattering of Light in Gases for $\bar{l} \geq \Lambda$

The classical hydrodynamic theory of the scattering of light ($\bar{l} \ll \Lambda$) was developed quite fully, in contrast to the theory of nonequilibrium phenomena giving rise to the scattering when $\bar{l} \geq \Lambda$, which began to develop only in the last decade. The stimulus was provided by investigations of the inelastic scattering of slow neutrons which is a convenient method for investigating molecular dynamics in condensed media [25-28].

When light of wavelength λ is scattered by a monatomic medium, the spectral distribution of the light scattered at an angle θ is proportional to $S(q\Omega)$, which represents a double Fourier transformation of the correlation function of the density $G(rt)$ ($\Omega/2\pi$ is the frequency shift of the scattered light and $q = 4\pi/\lambda \sin \theta/2$). In the limit of small values of q, the quantity $S(q\Omega)$ can be calculated from linearized hydrodynamic equations [29-30].

When fast processes are considered, the hydrodynamic equations should be replaced by the corresponding transport equations. The correlation function of the density in rarefied media can be investigated by means of the linearized Boltzmann equation [31]. This approach [31-35] provides an exact description of an ideal gas in the $\bar{l} \gg \Lambda$ limit and a correct hydrodynamic theory of the case when $\bar{l} \ll \Lambda$. The precision with which the linearized Boltzmann equation can be approximated in the intermediate region where $\bar{l} \sim \Lambda$ is comparable with the exact solution of the Boltzmann equation for Maxwellian molecules [36].

We shall consider the results obtained by Yip and Nelkin [33] who applied the transport model to the description of the time-dependent correlation function of the density representing the inelastic scattering of neutrons. Yip and Nelkin applied also this approach to the molecular scattering of light. They considered arbitrarily fast changes in space and time as a result of externally induced perturbations of the density and they assumed that the amplitudes of the oscillations in rarefied gases are small.

We have already mentioned the interest in the double Fourier transformation of the correlation function of the density which is given by

$$S(q\Omega) = \int\limits_{-\infty}^{\infty} dt \int d^3 r \, G(rt) \exp[i(qr - \Omega t)] \tag{I.19}$$

and which is related directly to the energy and differential cross section in the experiments involving the scattering of light. Figures 1 and 2 show graphically the solutions of the gas-kinetic equations in the adiabatic and isothermal ($\delta T = 0$) approximations.

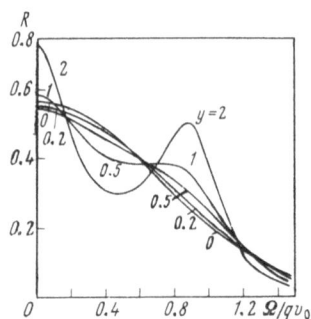

Fig. 1. Response of a gas to microscopic perturbations of the density, plotted for different
values of y. R = (qv_0/π) S(xy).

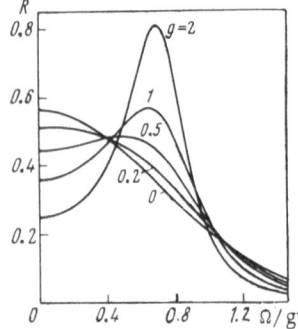

Fig. 2. Response of a gas to microscopic perturbations of the density, plotted for different
values of y. Isothermal approximation ($\delta T = 0$).

The parameter which represents the spectrum of the thermally scattered light is the
quantity proportional to the ratio of the wavelength of the Fourier component of the fluctuations
responsible for the scattering to the mean free path of molecules:

$$y = 2\pi/q/[(2\pi/\nu_{col})(2kT/m)^{1/2}] = \pi^{-3/2}\Lambda/\bar{l}, \qquad (I.20)$$

where ν_{col} is the effective collision frequency.

The principal characteristics of the correlation of the density contained in the Fourier
components of G(rt) are given in Figs. 1-3. When collisions are rare (i.e., y is small), all the
perturbations are absorbed by the free motion of the particles and since this motion is uncor-
related, the high-frequency components are absorbed more strongly.

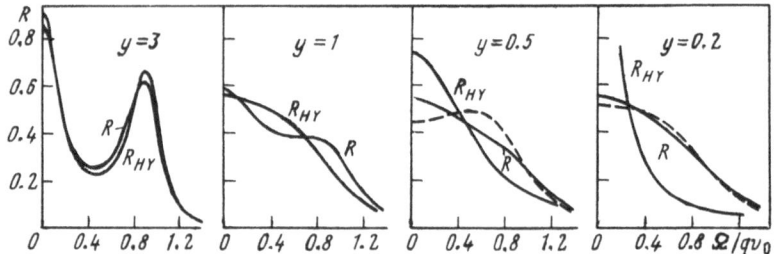

Fig. 3. Comparison of the gas-kinetic and hydrodynamic descriptions of the response of a gas to microscopic perturbations of the density, plotted in the form of R = $f(\Omega/qv_0)$. Here
R = (qv_0/π) S(xy) corresponds to the adiabatic approximation
and the gas-kinetic description; R_{HY} corresponds to the adiabatic approximation and the hydrodynamic description; the
dashed curves correspond to the isothermal approximation
and the gas-kinetic description.

In the collisionless limit we have

$$\lim_{y \to 0} S(xy) = [\pi^{1/2} q v_0]^{-1} \exp(-x^2), \qquad x = \Omega/q v_0. \qquad (I.21)$$

When the wavelength of the perturbations becomes comparable with the mean free path of the molecules, the correlation effects, because of consecutive collisions, become significant and this results in an appreciable reduction in the absorption at certain frequencies.

At high values of y these results can be understood quite easily if we consider the scattering of light by equilibrium fluctuations of the density. The processes in question result from fluctuations of the density and temperature (or entropy), as described in §1. The scattered-light spectrum consists of a central line, whose width is governed by the thermal diffusivity, and of a doublet which is symmetrical with respect to the center of the line and which appears as a result of scattering by elastic waves. The shift and widths of the doublet lines are governed by the frequency and absorption of sound with a given wave number q.

The intensity and the position of the scattered-light maximum are governed by the thermodynamic properties of the gas in question. Figure 4 compares the transport (gas-kinetic) and hydrodynamic descriptions of the Fourier transformation of the thermal correlation function $E(q\Omega)$ which represents the scattering of light at the undisplaced frequency. A comparison of the results obtained in the transport approach [33, 36, 37] with the results deduced in the hydrodynamic approximation [30, 31] (see also Fig. 3) shows that both descriptions are applicable if $y > 10$. The scattered-light spectrum consists of three separate nonoverlapping lines (the undisplaced central component and the symmetric doublet). In the $y \approx 5$ region these three lines begin to overlap since their widths increase faster with increasing q [Eqs. (I.7) and (I.11)] than does their splitting [Eq. (I.4)]. However, the whole scattered-light spectrum can still be described quite accurately by the hydrodynamic equations.

If $y < 2$, we must use the transport (gas-kinetic) theory. When the pressure in the gas is 1 atm each of those regions can be investigated by varying the scattering angle from 0 to 180°. This approach was suggested by Nelkin and Yip [38] and the appropriate experiments were carried out by Greytak and Benedek [39].

Greytak and Benedek used a frequency-stabilized single-mode laser, high-resolution Fabry—Perot scanning interferometers, and a photoelectric detection system. They measured the spectral distributions of light scattered in five gases (Ar, Xe, N_2, CO_2, CH_4) at 1 atm. The scattering angles were: 10.6° ($y \approx 13$), 22.9° ($y \approx 6$), and 169.4° ($y \leq 2$). Spectral measurements of the light scattered at an angle of 10.6° were used in the calculation of the velocity of hypersound in the gases being investigated. The ratio of the intensity of the Rayleigh line to the intensities of the Brillouin components was determined for CO_2 and Xe to within 3%. This ratio was in agreement with the Landau—Placzek relationship [1]

$$I_c/2I_B = (C_p/C_v - 1). \qquad (I.22)$$

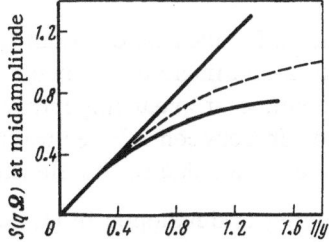

Fig. 4. Dependence, on $1/y$, of the width $S(q\Omega)$ at midamplitude. The continuous curve represents the exact solution of the Boltzmann equation for Maxwellian molecules; the dashed curve is the exact solution of the Navier—Stokes equation; the linear dependence is the solution of the linearized Navier—Stokes equation [30].

Greytak and Benedek used the spectrum observed at 22.9° to determine the half-width of the Brillouin doublet components for CO_2 and Xe and they estimated the phonon lifetime to be $(1.1-0.9) \times 10^{-8}$ sec.

The spectral distributions of the intensity of the scattered light obtained for the y < 2 case were in basic agreement with the predictions of the transport (gas-kinetic) theory [33].

§4. Stimulated Brillouin Scattering in Gases

In stimulated Brillouin scattering the intensity of the shifted components of the fine structure increases nonlinearly with the size of the scattering volume, measured in the direction of observation of the scattered light. This nonlinear rise of the Brillouin components occurs above a certain threshold value of the incident light intensity.

Chiao, Townes, and Stoicheff [40] were the first to observe stimulated Brillouin scattering in quartz and sapphire crystals. The same effect was observed in some liquids by Brewer and Rieckhoff [41, 42] and in amorphous solids by Mash, Starunov, Tiganov, Fabelinskii, and the present author [43].

Experimental determinations of the threshold intensity for stimulated Brillouin scattering indicated that this intensity varies considerably from one substance to another [44]. For example, the threshold for carbon disulfide is 30 MW/cm^2, whereas the threshold for benzene and some other liquids is $\sim$ 1200 MW/cm^2. The threshold of salol and triacetin is considerably higher [43].

The stimulated Brillouin scattering in compressed nitrogen, hydrogen, and oxygen was discovered by Mash, Starunov, Fabelinskii, and the present author [20]. Independently of us, this type of scattering was discovered by Rank et al. [21] and by Hagenlocker and Rado [22] in nitrogen, methane, carbon dioxide, and argon.

The physical origin of the stimulated Brillouin scattering is as follows. At high exciting radiation intensities the interaction between the excited and scattered light waves begins to exert a considerable influence on the nature of the motion in a medium, in contrast to thermal scattering in which this influence is unimportant and the motion can be described by Debye thermal waves. The application of an electric field to a body produces electrostriction which alters the volume of this body by an amount [1]

$$\Delta V/V = {}^1/_8 \pi \beta_s \left(\rho \frac{\partial \varepsilon}{\partial \rho} \right)_s E^2, \tag{I.23}$$

where β_s is the adiabatic compressibility.

If we use the definition $\beta_s = (1/\rho \; \partial\rho/\partial p)_s$, we can transform Eq. (I.23) into

$$|\Delta p| = p = {}^1/_8 \pi \left(\rho \frac{\partial \varepsilon}{\partial \rho} \right)_s E^2, \tag{I.24}$$

where $|\Delta p| = p$ is the excess pressure caused by electrostriction.

Approximate estimates [1] based on Eq. (I.24) show that a focused laser beam is capable of producing enormous pressure in a medium and, therefore, a nonlinear effect such as electrostriction (which is negligible under normal conditions) becomes of great importance. The stimulated scattering of light can be explained by the interaction between a laser wave, an elastic or pressure wave due to electrostriction, and a polarization wave due to the elastic wave.

The classical theory of the stimulated Brillouin scattering is developed in [1, 22, 45, 46].

Theoretical explanations of this phenomenon are given also in classical and quantum-mechanical treatments of the stimulated Raman scattering of light [47-50].

The stimulated Brillouin scattering resembles, in many respects, the Raman scattering of light. In an elementary scattering event a photon of laser frequency ω_L is absorbed, a photon of frequency $\omega_s = \omega_L - \Omega_{ac}$ is emitted, and an acoustic phonon $\hbar\Omega_{ac}$ is absorbed because of the strong attenuation of sound. When the absorption coefficient of hypersound α decreases gradually to a value comparable with the absorption coefficient of light K_ω, the nature of the scattering process changes. At high values of α the scattering is of the Raman type, in which the scattered light wave is mainly amplified, whereas at low values of α the process represents simultaneous parametric generation of acoustic and light waves.

Amplification of the Back-Scattered Stokes Component

The solution of the linearized wave equations for coherent acoustic and electromagnetic (exciting and scattered) waves yields an approximate expression for the energy gain experienced by the back-scattered Stokes component in the Brillouin process under steady-state conditions in the absence of saturation effects [1, 22].

Kroll drew attention to the fact that an allowance for transient processes was very important in the analysis of the stimulated Brillouin scattering. He analyzed in detail the scattering through 180° (back-scattering) in the case when the acoustic absorption is not very strong compared with the parametric gain [45]. If the lifetime of the acoustic phonons exceeds the duration of the laser pulse, the resultant amplification of the Stokes component should be less than that under steady-state conditions.

Hagenlocker et al. [51] varied the phonon lifetime by a factor of 10^2 by altering the temperature and pressure of various gases and they were thus able to investigate directly the effects of transient conditions on the stimulated Brillouin scattering. They showed that the gain calculated from the experimental data for the Stokes component of gases was several orders of magnitude smaller than the gain predicted by the steady-state theory but was in good quantitative agreement with the theoretical conclusions of Kroll [45].

It is known [52] that saturation occurs in one-photon resonance absorption and stimulated emission when the intensity of light reaches such a high value that the rate of resonance transitions becomes comparable with the rate of relaxation which maintains the difference between the populations of the initial and final states.

Similarly, saturation will occur in the stimulated Brillouin scattering if the stimulated scattering rate becomes sufficiently high.

Tang [46] discussed the stimulated Brillouin scattering under steady-state conditions and made allowance for saturation. In this case, the path in which an acoustic wave is absorbed is much shorter than the length of the interaction region or the distance in which the intensity of the exciting and back-scattered waves decreases significantly as a result of Brillouin scattering ($2\alpha^{-1} \leq x$). This situation is attained in Brillouin scattering experiments in gases in the 1-15 atm pressure range. When the acoustic wave is strongly attenuated the stimulated Brillouin scattering process can be described by two nonlinear first-order wave equations for the amplitudes of the Stokes component and of the exciting laser radiation [46]:

$$\frac{\partial E_s(x)}{\partial x} = \left[\frac{\mu\omega_s^2\gamma_1^2 k_p}{64\pi k_s k\left(i\Delta k - \frac{\alpha_p}{2}\right)}\right]|E_L(x)|^2 E_s(x) + \frac{\alpha_0}{2}E_s(x), \qquad (I.25)$$

$$\frac{\partial E_L(x)}{\partial x} = -\left[\frac{\mu\omega_L^2\gamma_1^2 k_p}{64\pi k_L k\left(i\Delta k + \frac{\alpha_p}{2}\right)}\right]|E_s(x)|^2 E_L(x) - \frac{\alpha_0}{2}E_L(x),\qquad(I.26)$$

where $\gamma_1 = P_1\varepsilon^2$; P_1 is the Pockels constant; k is the elastic constant; $k_p = \Omega^2\rho k^{-1}$; $k_s^2 = \omega_s\mu\varepsilon$; $k_L = \omega_L\varepsilon\mu$; $\Delta k = 1/v\,[\omega_s - \overline{\omega}_s]$; $\overline{\omega}_s = (\omega_L - \Omega)$ is the frequency of the Brillouin component which satisfies the Bragg condition of Eq. (I.1).

Equations (I.25) and (I.26), combined with the boundary conditions given in [46], describe the spatial variation of the amplitudes and phases of the exciting wave and the Stokes component.

If the Stokes component is amplified linearly, its intensity is a linear function of the initial intensity of the Brillouin components in the thermal scattering and an exponential function of the length of the interaction region and of the intensity of the laser radiation:

$$I_1 = I_{10}e^{gI_0L},\qquad(I.27)$$

$$g(\omega_s) = \frac{\mu\hbar\omega_L\omega_s\gamma_1^2 v^2\alpha_p}{2\varepsilon k\,[4(\omega_s - \overline{\omega}_s)^2 + \alpha_p^2 v^2]}\,.\qquad(I.28)$$

The threshold intensity of the exciting radiation corresponding to the onset of the amplification of the Stokes component is

$$\frac{|E_L|^2}{8\pi} \geqslant \frac{2\varepsilon k\alpha_0\,[\alpha_p^2 v^2 + 4(\omega_s - \overline{\omega}_s)]}{\gamma_1^2 k_s k_p \alpha_p v^2}\,.\qquad(I.29)$$

Equations (I.28) and (I.29) demonstrate that the maximum of the gain of the Stokes component and the minimum threshold for the appearance of the stimulated Brillouin scattering occur at the same frequency as the maximum of the spectral intensity in the thermal scattering ($\omega_s \equiv \overline{\omega}_s$).

The width and profile of the Brillouin scattering components are governed by the frequency dependence of the argument of the exponential function in Eq. (I.27) and by the spectral distribution of the intensity in the thermal scattering process. If $I_0gL \ll 1$, we find that $\delta\omega = 2\alpha_p v$, which is in agreement with the width of the Brillouin components observed in the thermal scattering (§1). If $I_0gL \gg 1$, the width of the Brillouin components is less:

$$\delta\omega = \alpha_p v\,[\ln 2/I_0gL]^{1/2}.\qquad(I.30)$$

An exact solution of Eqs. (I.25) and (I.26) provides a complete description of the amplification of the Stokes component with a suitable allowance for the saturation under linear and nonlinear conditions.

The theory of the stimulated Brillouin scattering in gases, like the theory of the thermal scattering, should be basically identical with the theory of the same phenomena in liquids provided the mean free path of the gas molecules remains much shorter than the wavelength of an elastic thermal wave which is active in the scattering at a given angle.

The expression for the threshold intensity of the exciting radiation $I_{th} \propto |E|_{th}^2$ for gases (I_{th}^{gas}) is formally of the same nature as the expression for the corresponding threshold in the case of liquids (I_{th}^{liq}). If we assume that the optical and acoustic losses in a liquid and in a compressed gas (~ 100 atm) are equal and the wave numbers in these media do not differ greatly, we find that

$$I_{th}^{gas} \approx 10^3 I_{th}^{liq}\qquad(I.31)$$

The results of experimental investigations of the stimulated Brillouin scattering in compressed gases [20-22, 53-57] demonstrate that:

1) the back-scattered radiation converges at an angle equal to the divergence angle of the exciting laser radiation;

2) the spectral width of the Brillouin scattering components is much less than the spectral width of the exciting radiation;

3) the time for the development of the Brillouin scattering can be very short (for example, 2 nsec) at the scattering threshold;

4) there is a definite relationship between the experimental conditions and the scattering efficiency: a given scattered power can be achieved by maintaining constant the value of the quantity $\rho P_0 \int_{l_1}^{l_2} dl/S$, where ρ is the density of the gas, P_0 is the power of the exciting radiation, and S is the cross section of the interaction volume along l (this relationship is satisfied well at low powers of the exciting radiation);

5) the velocity of hypersound calculated from Eq. (I.4) does not agree with the adiabatic value of Eq. (I.16) obtained at pressures 30-100 atm but it is close to the isothermal value.

§5. Stimulated Temperature (Entropy) Scattering of Light

The stimulated temperature (entropy) scattering is due to the interaction between a high-power laser wave and weak light waves resulting from the thermal scattering by fluctuations of the entropy in a nonlinear medium. Such scattering gives rise to temperature "waves" and to pumping of the energy of the laser radiation into the scattered light wave.

We shall now consider some aspects of the theory of the stimulated temperature (entropy) scattering [58], which is applicable in the hydrodynamic approximation ($\bar{l} \ll \Lambda$). We shall also estimate the possibility of observing this effect in gases. The stimulated temperature scattering was found in liquids by Zaitsev, Kyzylasov, Starunov, and Fabelinskii [59].

Starunov [58] assumes that the nonlinear interaction is due to the electrocaloric effect which alters the temperature of the medium (per unit time) by an amount [13]

$$\Delta \dot{T} = - T/4\pi \rho C_p \left(\frac{\partial \varepsilon}{\partial T} \right)_p \mathbf{E}\dot{\mathbf{E}}.$$

If the above expression is used in the heat conduction equation, we can find the law which governs the change in the temperature δT if we know the field E. When this equation is combined with the nonlinear Maxwell equations in which the nonlinear correction to the polarization of the medium is assumed to be $P^{NL} (\partial \varepsilon/\partial T) \delta TE$, the problem of the stimulated temperature scattering can be solved. It is assumed that three plane linearly polarized light waves travel in a medium so that the total intensity of the electric field is

$$E = \sum_{l=0}^{l=2} E_l \exp (i\omega_l - ik_l \mathbf{r}) + \text{complex conjugate} \tag{I.32}$$

The subscripts 0, 1, and 2 in Eq. (I.32) apply to the laser, Stokes, and anti-Stokes radiation, respectively; $\omega_0 = \omega_1 + \Omega = \omega_2 - \Omega$, $\Omega \ll \omega_0$. The calculation is carried out in the same way as in the solution of the problem of the stimulated scattering in the wing of the Rayleigh line [60].

Under steady-state conditions (i.e., when the time taken to establish a given temperature in a medium is shorter than the duration of the exciting radiation pulse), the heat conduction equation shows — after averaging over high frequencies — that the change in the temperature of the medium is

$$\delta T = \frac{-T\left(\frac{\partial \varepsilon}{\partial T}\right)_p}{16\pi\rho C_p}\left\{\frac{i\Omega E_0 E_1^*}{i\Omega + \chi(\mathbf{K}_0 - \mathbf{K}_1)^2}\exp i\Omega t - i(\mathbf{K}_0 - \mathbf{K}_1)\mathbf{r} + \right.$$

$$\left. + \frac{+ i\Omega E_0^* E_2}{i\Omega + \chi(\mathbf{K}_2 - \mathbf{K}_0)^2}\exp i\Omega t - i(\mathbf{K}_2 - \mathbf{K}_0)\mathbf{r} + \text{complex conjugate}\right\} \qquad (I.33)$$

Here, χ is the thermal diffusivity, ρ is the density of the medium, and C_p is its specific heat. It follows from Eq. (I.33) and from the Maxwell equations that the anti-Stokes component should decay and the Stokes component should grow in space exponentially:

$$|E_1(x)|^2 = |E_1(0)|^2 \exp\{g_{1T}x\}. \qquad (I.34)$$

Here, x is the coordinate in the direction of $\mathbf{K}_1$.

$$g_{1T} = -2k_\omega + B_T|\mathbf{K}_1|\frac{\Omega/\Omega_m}{1 + \Omega^2/\Omega_m^2}|E_0|^2, \qquad (I.35)$$

where $2k_\omega$ is the attenuation coefficient of light, and

$$B_T = \frac{T\left(\frac{\partial \varepsilon}{\partial T}\right)_p^2}{16\pi n^2 \rho C_p}, \qquad \Omega_m = \chi(\mathbf{K}_0 - \mathbf{K}_1)^2. \qquad (I.36)$$

The heating of the gaseous medium by the direct absorption of light can be ignored if

$$\left\{2nT\left(\frac{\partial n}{\partial T}\right)_p \chi(\mathbf{K}_0 - \mathbf{K}_1)^2/\varepsilon''\omega_0\right\} > 1, \qquad (I.37)$$

where ε'' is the imaginary component of the permittivity of the medium.

It follows from Eqs. (I.35) and (I.36) that the gain of the Stokes component in the stimulated temperature scattering has a maximum at the frequency $\omega_1 = \omega_0 = \Omega_m$, which corresponds to the half-width (measured at midamplitude) of the central line in the fine structure of the thermal scattering spectrum:

$$\Omega_m = \frac{\delta\omega_c}{2} = \chi(\mathbf{K}_0 - \mathbf{K}_1)^2 = \chi q^2. \qquad (I.38)$$

Numerical estimates obtained for benzene illuminated with unfocused laser radiation of $I \approx 100$ MW/cm^2 power density give $g_{1T} = 10^2$ cm^{-1} [59].

The maximum of the intensity in the stimulated temperature scattering spectrum is shifted by $\sim 10^{-3}$-10^{-4} cm^{-1} from the frequency of the exciting radiation and such a shift can be determined quantitatively only by the optical heterodyne method. In this method, the exciting line should have a half-width of 10^{-4}-10^{-5} cm^{-1}, whereas the usual half-width of the exciting radiation of the power indicated above is 10^{-2}-10^{-3} cm^{-1}. It follows that it would be difficult to determine the stimulated temperature scattering spectrum of liquids by the optical heterodyne method.

The situation is different in gases. For example, in the case of hydrogen at 2–10 atm, the half-width of the thermal scattering line resulting from isobaric fluctuations of the density (entropy) is $\sim 10^{-1}$-10^{-2} cm^{-1}. Therefore, if we use a spectroscopic instrument with a sufficiently high resolution (for example, a Fabry–Perot interferometer), we can – at least in principle – carry out a spectral investigation of the stimulated temperature scattering in gaseous hydrogen.

The condition of Eq. (I.35) is known to be satisfied for gases and the absorption of light can be ignored in the expression for the gain (I.35).

We shall now rewrite Eq. (I.35) so as to give the gain at the peak of the stimulated temperature scattering intensity, reduced to unit length and unit exciting radiation intensity:

$$g_{STS} = (g_{1T}/xI_L) = B_T | \mathbf{K_1} | \left(\frac{19}{300}\right)^2, \text{ cm/W}, \tag{I.39}$$

where the subscript STS refers to the stimulated temperature scattering. We shall compare the above equation with the corresponding expression for the gain in the case of the stimulated Brillouin scattering g_{SBS} (see §4).

We find that $g_{SBS} \approx$ (5–10) g_{STS} for hydrogen at 2–10 atm. An approximate relationship between the gain of Eq. (I.39) and the experimental threshold for the observation of the stimulated temperature scattering can be found by integrating dE_1/dx over the interaction region on the assumption that a plane laser radiation wave passes through a focusing lens. The integrated gain is then found to be equal to $\exp \{gP_0 l/a\}$, where P_0 is the laser radiation power and l/a is the ratio of the effective length to the cross-sectional area of the focal region.

If the laser radiation power is $P_0 \approx$ 100–150 MW and the ratio l/a is of the order of 10^4 [22], the integrated gain is found to be proportional to $\exp(15)$ for the stimulated Brillouin scattering and to $\exp(3)$ for the stimulated temperature scattering.

The minimum value of the gain necessary for the experimental observation of the stimulated scattering in gases is usually assumed to be $\exp(30)$ [22].

The estimates given above show that in order to observe the stimulated Brillouin or stimulated temperature scattering in gaseous hydrogen at 300°K at 2–10 atm, it is necessary to ensure an additional amplification of the scattered light. If the scattered light is observed at an angle of 180° with respect to the direction of the exciting radiation, the additional gain can be attained by regenerative amplification in the laser followed by amplification due to the four-photon interaction in a liquid with anisotropic molecules [61]. Moreover, it is also possible to achieve superregenerative amplification with the aid of the laser modes.

The method of trigger (superregenerative) amplification of the back-scattered light with the aid of the modes of the exciting source was used in the present investigation and will be described in the next chapter.

CHAPTER II

METHOD AND APPARATUS USED IN EXPERIMENTAL INVESTIGATIONS OF THE STIMULATED MOLECULAR SCATTERING OF LIGHT IN GASES

Our aim was to detect and determine the spectral distribution in the stimulated molecular scattering of light in gases. Therefore, the apparatus had to satisfy the following requirements:

1. The intensity of the optical radiation focused into the scattering region in a given substance should exceed the stimulated scattering threshold. For example, in

the case of stimulated Brillouin scattering, the threshold for gases is $I_{th}^{gas} \approx$ $10^3 I_{th}^{liq} \approx 10^4 \text{ MW/cm}^2$ (§4).

2. The optical part of the apparatus should make it possible to investigate the spectrum of the scattered light in which the dispersed components are shifted by $\sim 0.01 \text{ cm}^{-1}$ away from the exciting radiation frequency and extend over a spectral region ~ 0.01-3 cm^{-1}.

§6. Generation of High-Power Giant Ruby Laser Pulses with Narrow Emission Wavelengths

The first requirement was a high-power light source emitting in a narrow spectral range.

The present author [62] has already described a Q-switched ruby laser capable of emitting single pulses up to 300 MW power and of spectral width $\leq 0.01 \text{ cm}^{-1}$.

This laser was used in investigations of the stimulated Brillouin scattering spectra of gases [20, 53, 63]; in the detection and investigation of the stimulated wing of the Rayleigh line [64]; in studies of the stimulated Brillouin scattering in liquids, amorphous solids [43], and crystals at different temperatures [65]; and so on.

The applications of this laser obviously extend well beyond the purpose for which it was used in the present investigation. We shall now consider the construction and action of the laser.

Figure 5 shows schematically a composite Q-switched ruby laser used in the studies referred to above. Each of the two ruby crystals, 120 mm long and 10 mm in diameter (concentration of $Cr^{3+} \approx 0.05\%$), was placed in a separate enclosure consisting of two cylinders of elliptic cross section. One IFP-2000 flashlamp was placed at the focus of each ellipse and the ruby crystal was placed in the common focus of the two ellipses. The ruby crystals and the flashlamps were cooled with running water. The crystals were oriented so that their optic axes were parallel to each other. The laser radiation was polarized in the vertical plane.

All four flashlamps were connected in series and were fired simultaneously. They were supplied from a 600-μF capacitor bank, charged to 4-6 kV.

The two enclosures were arranged in series and placed in a resonator formed by a rear mirror (1) with a reflection coefficient $\sim 100\%$ and two plane-parallel plates made of fused quartz (6) and heavy flint glass (7), which acted as the front mirror of the resonator.

Giant pulses were excited by the "gain switching" method, in which an additional gain exceeding the resonator losses was introduced into the laser.

The Q factor of the resonator was switched by a saturable solution of cryptocyanine in methyl alcohol with a molar concentration of 1.8×10^{-6} M. This solution was poured into two plane-parallel cuvettes (2 and 4), 10 mm thick, placed between the mirror 1 and the ruby crystal 3 and between the ruby crystals 3 and 5.

The kinetics of the processes occurring in lasers with saturable filters of this type can

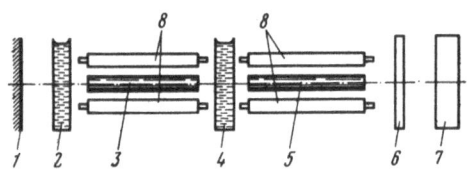

Fig. 5. Ruby laser: 1) plane mirror with a multilayer dielectric coating (R $\approx$ 100% at λ = 6943 Å); 2), 4) cuvettes containing solutions of cryptocyanine in methyl alcohol; 3), 5) ruby crystals; 6), 7) plane-parallel plates; 8) IFP-2000 flashlamps.

be described by means of rate equations whose solution for the case of self-excitation of the system is of the form [66]:

$$W_n n \geqslant W_m m + \tau_r^{-1},$$ (II.1)

where n and m are the differences between the populations of the Cr^{3+} ions in ruby and crypto-cyanine molecules in the solution, normalized to the total number N of the Cr^{3+} ions and M of the cryptocyanine molecules; W_n and W_m are the probabilities of stimulated transitions; τ_r is the decay time of the field in the resonator. It follows from this equation that, in principle, two excitation conditions can be realized in the laser described above.

The "soft" excitation occurred in the resonator I formed by the mirror 1 and the front end of the ruby crystal 3: in this case, a suitably selected pumping level produced such an initial value $n = n_0$ that the self-excitation condition of Eq. (II.1) was satisfied. In this case, the laser emission started from an intensity level governed by the spontaneous emission mechanism.

The "hard" excitation was obtained in the resonator II, formed by the mirror 1 and the plates 6 and 7. Since the resonator I was a component of the larger resonator II, an initial number of photons $\rho_0 > \rho_{th}$ was introduced at some moment into the resonator II, the solution in the cuvette 4 was bleached, and the system II reached the state given by Eq. (II.1).

The excitation conditions were selected so as to achieve high-power low-divergence radiation in a narrow spectral interval.

The use of saturable filters in Q switching reduced the width of the stimulated emission spectrum. This was supplemented by additional mode selection in passive interferometers to give single-mode emission (the output power under these conditions reached 5 MW [67]), whereas the usual width of the emission from ruby lasers with a rotating prism was of the order of $1 \, cm^{-1}$.

The Q factor of the laser described above was switched with the aid of two cuvettes (2 and 4) containing a solution of cryptocyanine in methyl alcohol. These cryptocyanine filters were bleached in a narrow spectral interval [68, 69] and they could reduce the emission spectrum to $1 \, cm^{-1}$ [70]. After bleaching of the solution in the cuvette 4, a short giant radiation pulse developed in the resonator 2. The spectral width of this pulse could be reduced by further mode selection with the aid of the plane-parallel plates 6 and 7 (these plates were 2.5 and 10 mm thick, respectively, and their refractive indices were 1.45 and 1.71). Moreover, the resonator III (Fig. 5) formed by the front end of the ruby crystal 3 and the plates 6 and 7 (this resonator contained the cuvette 4 and the ruby crystal 5) acted as an asymmetric regenerative amplifier [71] in which feedback was modulated by the saturable filter 4. This amplifier was used as an active interference filter which participated directly in the formation of a giant pulse. In this case, the following conditions should be satisfied [72]:

$$R_1 G^1 > 1, \quad R_2 < R_1, \quad R_0 G^1 < 1,$$ (II.2)

where R_1 is the reflection coefficient of the plates 6 and 7; R_2 is the reflection coefficient of the front end of the ruby crystal 3; $R_0 = (R_2 R_1)^{1/2}$; G^1 is the gain per pass; $R_0 G^1 = 1$ is the condition for stimulated emission in the resonator III. To achieve these conditions, the ruby crystal 5 should be detuned relative to the crystal 3 and to the plane-parallel plates 6 and 7. The active asymmetric interference filter operated in the reflection and transmission pass band:

$$\Delta v = \frac{[1 - R_0 G^1]}{2\pi (n_1 d + n_2 l)(R_0 G^1)^{1/2}} \left[1 + \frac{(1 + R_0 G^1)^2}{6 R_0 G^1} \right],$$ (II.3)

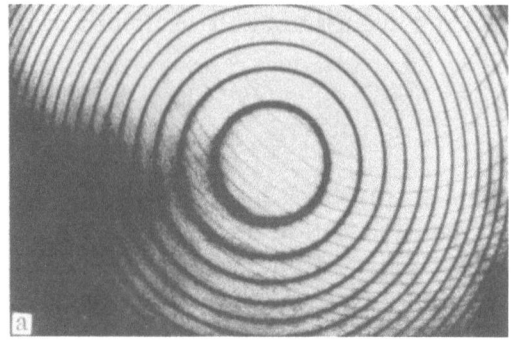
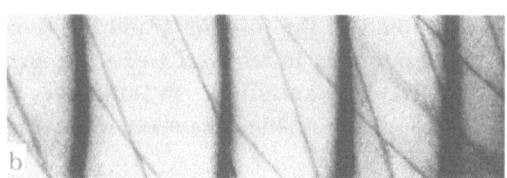

Fig. 6. Interferograms of laser radiation recorded for $\Delta\nu_{FP} = 0.166$ (a) and 1.5 cm^{-1} (b).

where n_1 is the refractive index of ruby; n_2 is the refractive index of air; d is the length of the ruby crystal; l is the length of the gap between the ruby and the reflecting elements in the resonator III.

Initially, when the gain per pass was high, we found that $\Delta\nu \leq 0.01$ cm^{-1}. When the conditions of Eq. (II.2) were satisfied, the gain resulting from the reflection in the active interference filter could be higher than the transmission gain and this would reduce the time necessary for the development of a giant pulse [73].

Equation (II.3) was derived without allowance for saturation. The various modes are amplified independently of one another as long as their total power is below the value sufficient to reduce the population inversion. The amplification process then becomes nonlinear and the emission spectrum may become narrower because of mode competition [74]. It is shown in [75] that the difference between the losses and the gain, needed to suppress one mode relative to another, is inversely proportional to the number of passes necessary to reach maximum gain. In the laser described above, this difference is usually achieved after several hundreds to several thousands of passes because the stimulated emission is excited under "soft" conditions, whereas in rapidly switched lasers (those modulated by a Kerr prism or cell) the number of passes required for this purpose is of the order of ten.

The laser described in the present section emitted giant single pulses of 10-12 nsec duration and up to 300 MW power (the power depended strongly on the quality of the ruby crystals and, particularly, on the quality of the crystal 3). The width of the emission spectrum was ≤ 0.01 cm^{-1} and the divergence of the radiation was $2\varphi \approx 20\text{-}30'$.

Figure 6 shows typical interferograms of the laser radiation obtained in different spectral ranges ($\Delta\nu_{FP}$) of a Fabry–Perot interferometer.

The oscillograms of giant pulses (Fig. 7) indicated that the pulse shape depended on the pumping energy provided by the flashlamps. Interferograms and oscillograms obtained simultaneously (Fig. 7) and calorimetric measurements of the laser radiation showed that, when the threshold was exceeded by 50%, the width of the emission spectrum was 0.005-0.01 cm^{-1} and the output power was constant to within 10%.

The output power of the laser could be varied continuously within wide limits, retaining the narrow emission spectrum (≤ 0.01 cm^{-1}). This could be done by varying the concentration of the cryptocyanine solution in the cuvettes 2 and 4 and by altering suitably the pumping energy.

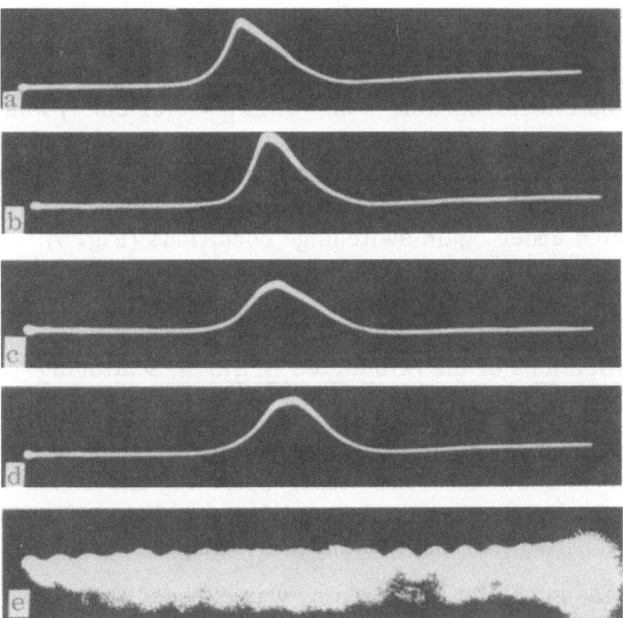

Fig. 7. Oscillograms of a giant pulse emitted by a laser under "gain switching" conditions. Pumping energy (J): a) 16,000; b) 14,700; c) 12,000; d) 11,000. S1-14 oscillograph, scanning time 100 nsec, timing marks at 8-nsec intervals (e), triggering provided by an FÉU-15 photomultiplier, signal provided by an FÉK-09 photocell.

We also studied a conventional laser (Fig. 8) in which the Q factor was switched by a single saturable filter (Fig. 8).

Figure 9 shows interferograms of the radiation emitted by the laser operating under the "gain switching" conditions (a and b) and under the standard Q-switching conditions (c and d)

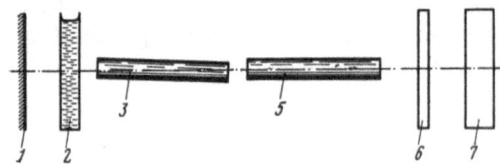

Fig. 8. Laser with one saturable filter. The components are the same as in Fig. 5.

Fig. 9. Interferograms of laser radiation obtained under different emission conditions (Figs. 5 and 8): a), c) laser threshold; b), d) 50% excess of the pumping power over the threshold. Spectral range of the interferometer 0.166 cm^{-1}.

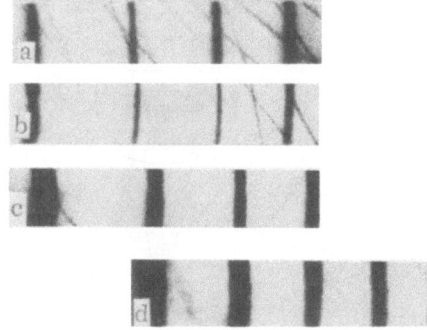

These interferograms were obtained either at the laser threshold (a and c) or for the same value of the excess ($\sim 50\%$) over the threshold (b and d).

The line width in the interferograms a and b ($\Delta\nu_L = 0.01$ cm^{-1}) was governed by the transfer function of the interferometer [76], whereas under normal Q-switching conditions (c and d) the line was much broader when the laser threshold was exceeded by $\sim 50\%$ ($\Delta\nu_L \approx 0.05-0.1$ cm^{-1}) (Fig. 9d). The output power of the laser shown in Fig. 8 was only about half the output power of the laser operated under "gain switching" conditions (Fig. 5).

§7. Optical Components of the Apparatus

Apparatus Used in Investigations of the Stimulated Brillouin Scattering in Compressed Gases Under Integrating Conditions

The stimulated Brillouin scattering in gases was investigated using apparatus shown schematically in Fig. 10. A Q-switched ruby laser operating under the "gain-switching" conditions (§6) was used as the light source. The ruby laser radiation was focused by an $f = 3$ cm lens inside a gas chamber filled with the gas under investigation.

The arrangement was such that the detector was exposed to light which was scattered at 0° with respect to the direction of propagation of the exciting radiation and to the back-scattered light which passed through the laser and returned to the chamber V after reflection from a mirror M. Moreover, some of the laser radiation could also reach the detector.

The radiation emerging from the chamber V was directed by a glass plate SP$_2$ through a filter F to a Fabry−Perot interferometer F-P. The spectrum produced by the interferometer was photographed on a photographic plate PL. The focal length of the camera objective was L$_2$ = 120 or 160 cm.

A second splitter SP$_1$ was used to photograph the back-scattered light (Fig. 10). However, in this case, the reflection losses at the SP$_1$ plate reduced the intensity of the initial exciting radiation. Therefore, the first experiments were carried out with the SP$_1$ plate removed and the radiation was reflected from the SP$_2$ plate. The stimulated Brillouin scattering was investigated using interferometers with spectral ranges of 0.166, 0.333, 1, and 0.07 cm^{-1}.

In the alignment of the laser, the mirror M, the plane-parallel plates m$_1$ and m$_2$, and the ends of the ruby crystal R$_1$ were made parallel to one another by an autocollimator taken from the optical bench set of the OKS-2 type. The second ruby crystal (R$_2$) was deflected through about 4' in such a way that the normal to its end surface remained parallel to the plane of polarization of the output radiation. The laser radiation spectrum was then photographed after

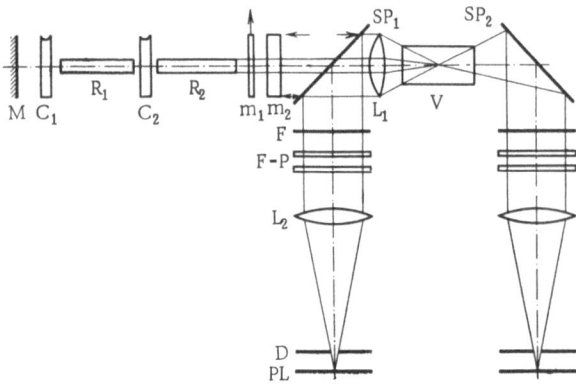

Fig. 10. Apparatus used in investigations of the stimulated Brillouin scattering. M is a mirror ($R \approx 100\%$); C$_1$ and C$_2$ are cuvettes, 1 cm thick, containing a solution of cryptocyanine in methyl alcohol of $1.8 \times 10^{-6} M$ concentration; R$_1$ and R$_2$ are ruby crystals, 12 cm long; m$_1$ and m$_2$ are plane-parallel plates; SP$_1$ and SP$_2$ are glass beam splitters; L$_1$ is a lens of $f = 3$ cm focal length; L$_2$ is an objective ($f = 120$ cm); V is a high-pressure gas chamber; F-P is a Fabry−Perot interferometer; PL is a photographic plate.

mode selection. Finally, the stimulated scattering spectra were photographed. A check on the stability of the conditions was provided by photographing the laser radiation spectrum before and after recording the stimulated scattering spectrum.

Apparatus Used in Investigations of the Time Evolution of the Stimulated Brillouin Scattering in Gases

The evolution of the stimulated Brillouin scattering in time was investigated with a fast-scanning image converter [53, 77].

The apparatus is shown schematically in Fig. 11. It could be used to record simultaneously the time dependence of the integrated intensity of the transmitted and back-scattered radiation (this could be done with the aid of FÉK-09 photodiodes). A Fabry—Perot interferometer and a PIM-3 image converter made it possible to obtain integrated and time-scanned spectrograms.

The exciting radiation was in the form of giant pulses emitted by a ruby laser which was Q-switched by a Pockels cell. This Q-switching method was selected because of the convenience of triggering the electronic circuit of the image converter by the pulse used to start the Pockels cell. The investigation was carried out at the laser emission threshold. Under these conditions, the laser produced giant single pulses of about 20 nsec duration, 40-50 MW power, and 0.005-0.01 cm^{-1} spectral width.

The laser radiation was focused by an f = 3 cm lens inside a chamber filled with nitrogen kept at a pressure of 150 atm. Some of the light scattered at angles of 180° and 0° (with respect to the direction of propagation of the exciting radiation) was split off by glass plates m_3 and m_4 and recorded oscillographically with the aid of the photodiodes Fd_2, Fd_1, and an S1-11 oscillograph. Silvered mirrors (m_5, m_6, and m_7) were used to deflect the major part of the scattered light onto Fabry—Perot interferometers $F-P_1$ and $F-P_2$ and a calorimeter C. The interference pattern produced by the $F-P_1$ interferometer with the aid of an objective L_2 (f = 80 cm) was focused onto the slit (30 μ wide) of a PIM-3 image converter (IC in Fig. 11). A time scan of the interference rings, which appeared on the image-converter screen, was photographed.

The time resolution was governed by the time needed for the formation of the interference pattern and amounted to 1.5-2 nsec.

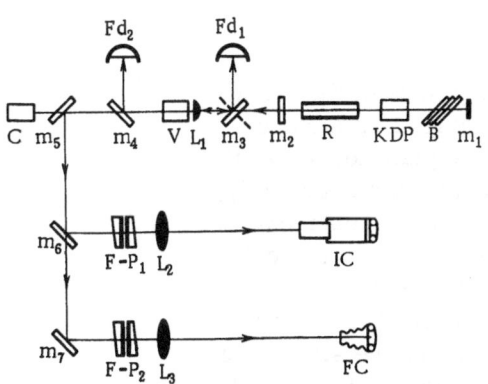

Fig. 11. Apparatus used in investigations of the time evolution of the stimulated Brillouin scattering. Here, m_1 is a dielectric mirror (R $\approx$ 100%); B is a Brewster pile; KDP is a Pockels cell made of KDP; R is a ruby rod, 120 mm long and 12 mm in diameter; m_2 is a dielectric mirror (R $\approx$ 30%); m_3 and m_4 are plane-parallel glass plates; Fd_1 and Fd_2 are photodiodes; V is a gas-filled chamber; L_1 is a lens of f = 3 cm focal length; m_5, m_6, and m_7 are silvered mirrors (R $\sim$ 80, 50, and 96%, respectively); C is a calorimeter; $F-P_1$ and $F-P_2$ are Fabry—Perot interferometers with a spectral range 0.166 cm^{-1}; L_2 and L_3 are objectives of 80- and 70-cm focal lengths, respectively; IC is an image converter; FC is a camera.

Neutral filters were used to ensure that the intensity of the scattered light was in the linear response region of the photodiodes and the image converter.

The interference pattern produced by the second interferometer (F-P$_2$) was photographed in the integrated form by an objective (f = 70 cm) and a camera FC. The effective optical path between the focus of the lens L$_1$ and the 100% reflecting mirror m$_1$ was ~105 cm.

Apparatus Used in Spectral Investigations of the Stimulated Brillouin Scattering and Stimulated Temperature Scattering in Low-Pressure Gases

The results of studies of the stimulated molecular scattering of light in gases were obtained using apparatus shown schematically in Figs. 10 and 11. These results and theoretical treatments made it possible to solve some of the problems relating to the methodology of detection of the stimulated scattering and to investigate ways for further studies of the spectral composition of the scattered light in the case of nonlinear interaction between high-power laser radiation and gaseous media.

The apparatus shown schematically in Fig. 12 was used to detect the stimulated Brillouin and stimulated temperature scattering in gaseous hydrogen at low pressures. The anti-Stokes broadening of the laser radiation spectrum transmitted by the interaction region in various gases was observed. The stimulated Brillouin scattering in compressed nitrogen was studied as a function of the intensity of exciting radiation using various detection systems.

Studies of the stimulated Brillouin and the stimulated temperature scattering in low-pressure gaseous hydrogen were carried out under conditions in which feedback occurred between the scattering region and the laser, i.e., the quarter-wave plate was not used.

The stimulated molecular scattering in low-pressure gases could be detected only because of the additional superregenerative amplification of the light scattered through θ = 180° by the laser modes (§8).

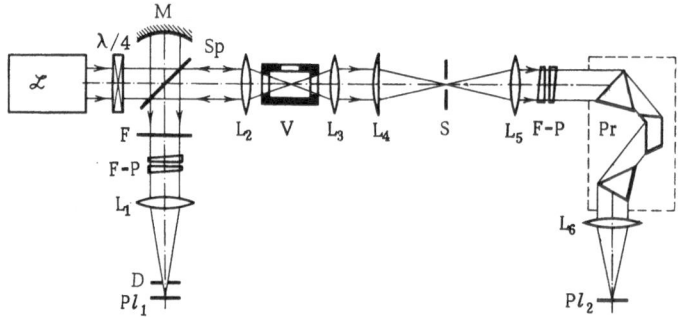

Fig. 12. Apparatus used in investigations of the stimulated Brillouin temperature scattering. L is a laser (Fig. 5); $\lambda/4$ is a mica quarter-wave plate for λ = 6943 Å radiation; M is a spherical mirror (f = 6 m, R $\approx$ 96%); Sp is a glass beam splitter; L$_2$ and L$_3$ are spherical lenses; V is a high-pressure gas chamber; L$_4$ is a cylindrical lens (f = 10 cm); F-P is a Fabry–Perot interferometer; L$_1$ is an objective (f = 160 cm); F is a red filter; D is a diaphragm; S, L$_5$, and Pr are the slit, objective, and prism of an ISP-51 spectrograph; L$_6$ is an objective (f = 80 cm); Pl_1 and Pl_2 are photographic plates.

The exciting radiation was in the form of giant pulses generated by a ruby laser L, described in §6. Eight IFP-2000 flashlamps (Fig. 5) were used to provide a wider range of pumping powers. Each of the two ruby crystals was placed in a separate enclosure consisting of four cylinders of elliptic cross section.

Giant laser pulses of 50-150 MW power were produced by varying the initial transmission of the saturable filters, used for the switching of the Q factor of the resonator, and by selecting the appropriate pumping power level. These pulses were spectrally narrow, $(0.5-1) \times 10^{-2}$ cm^{-1}, and were obtained for 50% excess of the pumping level over the laser threshold. The actual investigation was carried out at pumping levels 25-30% in excess of the threshold.

The laser radiation was focused by an $f = 20$ cm lens inside a chamber 10 cm long. This chamber was filled with gaseous hydrogen. The power of the incident radiation was selected so as to avoid the generation of a plasma in the focusing region. Plasma would not only absorb the exciting radiation but could also give rise to undesirable reflections [78]. The appearance of plasma could be monitored visually through a side window in the chamber. The light at the displaced frequency was scattered back into the laser ($\theta = 180°$) and it excited that resonator mode which was nearest in frequency to the peak of the stimulated scattering gain. A glass beam splitter Sp and a spherical mirror M ($f = 6$ m) were used to direct some of the light through a filter F to a Fabry–Perot interferometer F-P and this light was photographed on a plate Pl$_1$.

The Fabry–Perot etalons had spectral ranges of 0.166, 0.32, and 0.5 cm^{-1}.

We used the same optical apparatus in a spectral investigation of the stimulated Brillouin scattering in nitrogen kept at a pressure of 125 atm. This study was carried out at different intensities of the exciting radiation in the interaction region. The laser was operated at its stimulated emission threshold in the presence or absence of feedback between the scattering region and the laser. In the former case, several components of the Brillouin scattering appeared in succession, whereas, in the latter case, the scattering region and the laser were separated by a quarter-wave plate.

When the ruby laser radiation, polarized linearly along the z axis, passed through the quarter-wave plate ($\lambda/4$ in Fig. 12), its polarization became circular. The back-scattered radiation also had circular polarization. Light of circular polarization transmitted by the quarter-wave plate was converted into light polarized linearly along the y axis. Such light could not be amplified in the ruby laser. The most effective decoupling between the scattering region and the laser was achieved when the plane-parallel cuvette 4 of Fig. 5 was inclined at the Brewster angle with respect to the exciting beam.

The quarter-wave plate was aligned as follows. A rough position was determined with the aid of a $\lambda = 6328$ Å gas laser. Next, the stimulated Brillouin scattering was excited in liquid CS$_2$, known to have a low threshold for the appearance of the Brillouin component, compared with gases. Slight rotation of the quarter-wave plate relative to the optic axis of the laser was used to ensure that only one Stokes component of the Brillouin scattering appeared in the interferogram recorded by the Pl$_1$ plate. Finally, the cell containing CS$_2$ was replaced by the gas-filled chamber.

In a spectral investigation of the light which traversed the focusing region we found that the laser radiation spectrum experienced anti-Stokes broadening in gases kept at pressures of 1-150 atm. In these experiments, a 150-MW light pulse was focused by a lens ($f = 3$ cm) inside a chamber filled with gaseous nitrogen or hydrogen. A spherical lens L$_3$ ($f = 3$ cm) and a cylindrical lens L$_4$ were used to focus the light transmitted by the gas-filled chamber onto the slit of an ISP-51 spectrograph. A Fabry–Perot interferometer (spectral range 0.6 or 1.66 cm$^{-1)}$ was crossed with a prism spectrograph in an arrangement in which an internal parallel beam was used [76].

§8. Mechanism of Trigger Amplification of the Stimulated Brillouin Scattering and Stimulated Temperature Scattering Components

The apparatus described in the preceding sections can be used to amplify considerably the light scattered back into the laser. If the time taken by the scattered light to return to the laser is considerably shorter than the duration of the laser pulse, the scattered light is amplified since its frequency is within the limits of the width of the R_1 luminescence line of ruby. This amplified component of the stimulated scattering can be sufficiently strong to cause further stimulated scattering, thus acting like a second laser mode.

Brewer [79] investigated the stimulated Brillouin scattering in liquids and found that the scattered Stokes component was amplified in the laser under superregenerative conditions, i.e., the axial mode nearest in frequency to the Brillouin scattering component was excited. Under these conditions, the precision of the shift of the Brillouin scattering components recorded in an interferogram was governed by the separation between the axial modes of the laser ($\Delta\nu_m = c/2L$, where L is the optical length of the resonator).

This mechanism of the amplification of the stimulated scattering components can be called the "trigger mechanism" since weak scattered light is sufficient to switch the laser to the biharmonic operation because of feedback with the scattering region.

To understand the effective amplification of the scattered radiation by the laser modes, we must analyze the operation of a laser whose emission line is sufficiently wide to encompass many oscillation modes. Basov, Morozov, and Oraevskii [74] derived equations for a laser emitting two oscillation modes within the width of the emission line of ruby and they investigated the cases of strong and weak detuning of the modes relative to one another, making allowance for their nonlinear interaction.

We shall be interested in the case when the natural frequencies of the oscillation modes are separated from one another by an interval sufficient to ensure that their resonance curves do not overlap, i.e., that the losses in the system have separate minima at frequencies ω_1 and ω_2. In a plane-parallel resonator these are the modes with different axial indices. An analysis of the equations of Basov et al. [74] shows that, depending on the negative losses at frequencies ω_1 and ω_2, oscillations can appear at both frequencies simultaneously (this corresponds to the simultaneous excitation of several oscillation modes) or at one frequency only (one mode only is excited). Moreover, in either case the steady-state conditions are stable.

The diagram shown in Fig. 13 can be used to determine which steady state is attained by a system for a given excess of the pumping power over the threshold and a given Q factor of the oscillation modes (Q_i). The coordinate axes in Fig. 13 represent the ratios of the losses for each mode to the gain, $P_i = h_i \gamma_i^{-1}/k$, where $h_i = \omega_i/2\omega_1 Q_i$; the coefficients γ_i represent the line profile; μ_{12} depends on the degree of overlap of the fields of different modes; for a plane-

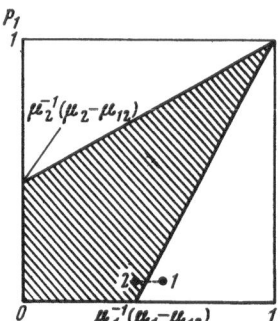

Fig. 13. Diagram for the determination of the steady-state operation of a laser emitting two modes under strong detuning conditions.

parallel resonator we have $\mu_1 = \mu_2$ and the ratio μ_{12}/μ_1 assumes the values 8/27, 4/9, and 2/3, depending on the number of identical indices representing the distribution of the fields.

If a point (P_1, P_2) lies within the shaded region, both oscillation modes are excited and are both stable. If this point lies outside the shaded region, the stable conditions are those in which the mode with the lower losses is excited.

Consequently, when radiation emitted by a laser generating a single mode ω_1 (point 1 in Fig. 13) is focused into a scattering medium, the stimulated back-scattered light at the shifted frequency ω_2 can excite a laser mode which is nearest to it in respect of the frequency (because of an increase in the Q factor Q_2 or, which is equivalent, because of an increase in the gain of the mode of frequency ω_2) and the laser emission becomes biharmonic (point 2 in Fig. 13).

This process can be controlled by suitable selection of the pumping level and of the Q factors of the oscillation modes in a certain part of the spectrum near the R_1 ruby line. The number of modes participating in the amplification of the stimulated scattering depends on the width of the scattered radiation line and on the spectral separation between the axial laser modes which can be fitted into the width of the emission line of ruby. Thus, the width of the components of the stimulated Brillouin scattering in gases at ~ 100 atm or higher pressures should be less [Eq. (I.30)] than the width of the fine structure in the case of thermal scattering [Eq. (I.7)], i.e., $\delta\nu_{SBS} < \delta\nu_{TS} \leq 0.005$ cm^{-1}. This value corresponds approximately to the width of a mode in the exciting radiation but it can be less than the spectral separation between the modes. In the latter case the amplification will occur at only one laser mode. The width of the components will increase with decreasing pressure and, consequently, several modes may take part in the amplification of light scattered in low-pressure gases.

We shall now consider how the amplification of the stimulated scattering occurs in the laser modes if the exciting radiation is in the form of pulses produced by a ruby laser operated under "gain switching" conditions (§6).

We have already mentioned that this particular laser can produce giant pulses of 10-12 nsec duration, up to 300 MW power, and $\Delta\nu \leq 0.01$ cm^{-1} spectral width. When the pumping power is 50% higher than the laser threshold, the width of the emission spectrum remains practically constant (Fig. 9) because of the special nature of the "gain switching" operation. One of the components used in the "gain switching" arrangement is an asymmetric regenerative amplifier which is used as an active interference filter with reflection and transmission pass band $\Delta\nu_F < 0.01$ cm^{-1}.

Clearly, this regenerative amplifier will behave as an active interference filter, with the same transmission band, in respect of the back-scattered light.

The spectral separation between the neighboring axial modes of the laser is 0.005 cm^{-1}.

If we select the optimal experimental geometry in which the back-scattered light reaches the regenerative amplifier and the gain per pass is high (i.e., $\Delta\nu_F < 0.01$ cm^{-1}), the stimulated scattering component is amplified in just one laser mode. This enhances considerably the precision of measurements of the components of the stimulated scattering in low-pressure gases. In this case, the precision of the determination of the shifts of the components is limited solely by the spectral separation between the axial laser modes (0.005 cm^{-1}): in the case of the fine structure observed for hydrogen, this corresponds to 2-3%.

The numerical estimates given in §4 show that when the exciting radiation is of 100-150 MW power the exponential gain of the Stokes component of the stimulated Brillouin scattering in hydrogen at 2-10 atm is 5-10 times greater than the gain of the stimulated temperature (entropy) scattering. Moreover, it follows from Eqs. (I.3) and (I.11) that the components of the stimulated Brillouin and temperature scattering are shifted by different amounts relative to the laser line. For example, in the case of hydrogen at 6 atm, the shift of the Brillouin scattering components is $\Delta\nu_B \approx 0.14$ cm^{-1}, whereas the shift of the temperature scattering components is $\Delta\nu_T \approx 0.14$ cm^{-1} (this shift corresponds to the half-width of the thermal scattering line resulting from fluctuations in the entropy; see also Fig. 1).

Consequently, the components of the stimulated Brillouin and temperature scattering may be amplified simultaneously in the laser modes: it all depends on the strength of the feedback, the pumping power, and the Q factors of the oscillation modes, whose frequencies correspond to the gain maxima of the stimulated Brillouin and temperature scattering in the interaction region. This amplification will occur independently for the two types of scattering until the total power of the exciting laser mode ω_L and of the two other modes (ω_B and ω_T) reaches a value sufficient for reduction of the population inversion in ruby. When this happens, competition between the modes may occur [74, 75].

Such mode competition was observed experimentally (Fig. 14e). In this case, two plane-parallel fused-quartz plates, 2.5 mm thick and separated by a ring 5 mm thick, were used to separate the laser modes.

The stimulated thermal scattering was investigated by separating the modes in such a way that the Q factor of those modes whose frequency coincided with the frequency of the components of the Brillouin scattering was less compared with the Q factor of the modes which coincided in frequency with the maximum of the stimulated temperature scattering (the separator consisted of two quartz plates, 13 and 2.5 mm thick, held apart by a 5-mm-thick ring). Spectrograms of the stimulated temperature scattering amplified in the laser modes are shown in Figs. 14b and 14d.

In both cases, we obtained interferograms which contained components of the stimulated Brillouin and temperature scattering.

The Stokes component of the Brillouin scattering amplified in a laser mode could excite a new Stokes component separated from the undisplaced laser line by twice the usual distance. This component could be amplified in another laser mode, etc. (Fig. 14f). This regenerative

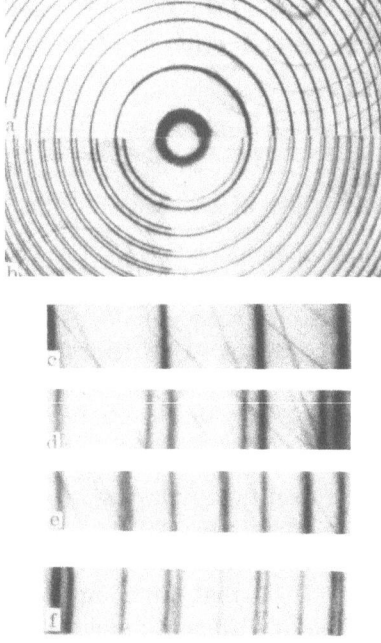

Fig. 14. Spectra of stimulated Brillouin and temperature scattering in hydrogen, amplified by laser modes: a) laser radiation ω_L; b) ω_L and a Stokes component of the stimulated temperature scattering (H$_2$ pressure 6 atm, spectral range of Fabry–Perot interferometer $\Delta\nu_{FP} = 0.166$ cm^{-1}); c) laser radiation ω_L; d) ω_L and a Stokes component of the stimulated temperature scattering; e) ω_L and a Stokes component of the stimulated Brillouin scattering; f) ω_L and two Stokes components of the stimulated Brillouin scattering (H$_2$ pressure 3.5 atm, spectral range $\Delta\nu_{FP} = 0.312$ cm^{-1}).

amplification mechanism gave rise to several equidistant lines in the interferograms. The number of such lines depended on the experimental geometry and on the duration and intensity of the exciting light pulses. This was also true of the stimulated temperature scattering if the maximum of this scattering was amplified in one laser mode.

Control experiments on the trigger amplification of the components of the Brillouin and temperature scattering indicated that, at 30-40% excess of the pumping power over the threshold value, the laser was stable only during 50-70 flashes, as judged by the reproducibility of the results obtained in a study of the spectral composition of the stimulated molecular scattering of light in gases. The instability was mainly due to changes in the initial transmission of the saturable filters used in Q switching of the laser.

Stable operating conditions were achieved in the laser before each series of experiments.

The interferograms shown in Figs. 14a and 14b were obtained by photographing ten flashes in succession on the same film. These flashes were separated by intervals of 3 min and the aperture of the diaphragm D was selected in a suitable manner and different exposures were used.

CHAPTER III

EXPERIMENTAL RESULTS OF SPECTROSCOPIC INVESTIGATIONS OF THE STIMULATED BRILLOUIN SCATTERING IN GASES

§9. Investigations of Interferometric Spectra of the Stimulated Brillouin Scattering in Compressed Gases

We used the classical approach and the data on the absorption of sound in gases to demonstrate (§2) that discrete components of the fine structure of the Rayleigh line should be observed in compressed gases. The condition for the existence of these discrete components is the inequality (I.8): $\alpha \Lambda \ll \pi$. Our calculations showed (§2) that $\alpha \Lambda = A\bar{l}/\Lambda$ for gases. At atmospheric pressure ($l \sim 10^{-5}$ cm, $\Lambda \sim 3 \times 10^5$ cm) the discrete structure cannot be observed in the thermal scattering because $\alpha \Lambda > \pi$ but this is not true of pressures of 20-30 atm ($\alpha \Lambda < \pi$).

If the theoretical conclusions are correct, we should observe the discrete fine structure of the Rayleigh line in compressed gases (~ 20 atm) and, in principle, the stimulated Brillouin scattering should also be observed. If the condition of $\bar{l} < \Lambda$ is satisfied, the expressions for the threshold of the stimulated Brillouin scattering in gases and liquids are formally identical, but at pressures of 100 atm the threshold for gases is 10^3-10^4 times higher than for liquids (§4).

The stimulated Brillouin scattering in compressed gases was investigated by the present author employing apparatus described in §7. The output power of the laser was 250 MW. The scattering was studied in nitrogen at 100 and 125 atm, in oxygen at 75, 100, and 150 atm, and in hydrogen at 95 atm.

The spectra of nitrogen and oxygen had four Stokes components and, occasionally, one weak anti-Stokes component. The spectrum of hydrogen comprised two Stokes components. Figure 15 shows typical interferograms of the stimulated Brillouin scattering in compressed gases.

The stimulated Brillouin scattering was not observed in helium at 140 atm. This was expected because, under the experimental conditions employed, the scattering threshold for

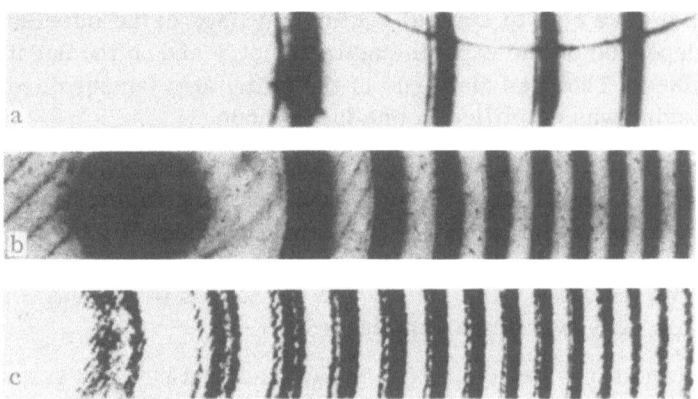

-Fig. 15. Spectra of the stimulated Brillouin scattering in
compressed gases: a) hydrogen (P = 95 atm), spectral
range of Fabry–Perot interferometer $\Delta\nu_{FP}$ = 1 cm^{-1};
b) oxygen (P = 150 atm), $\Delta\nu_{FP}$ = 0.333 cm^{-1}; c) nitrogen
(P = 125 atm), $\Delta\nu_{FP}$ = 0.166 cm^{-1}.

helium was more than one order of magnitude higher than the thresholds for the other gases
[see Eq. (I.29) and Table 1].

The positions of the discrete components in the Brillouin spectrum were used to determine the velocity of hypersound in all three gases. The results of this determination are given
in Table 1 alongside the adiabatic and isothermal velocities of sound. Other properties of the
gases and the experimental conditions are also included in this table. The results obtained
showed that the velocity of hypersound in nitrogen and hydrogen was less than the adiabatic
velocity of sound in these gases although one would expect the velocity of hypersound to be considerably higher than the adiabatic value v_0 because of relaxation and potential heating in the
scattering region.

The average velocity of hypersound in oxygen was close to the adiabatic value but detailed measurements showed that the shift of the first Stokes component was smaller than that of
the second component, and that of the second was smaller than that of the third. The velocity
of hypersound calculated from the first component was close to the isothermal value. It was
possible that the observed reduction in the velocity was due to the occurrence of the stimulated
Brillouin scattering in a nonequilibrium plasma, generated in the focused beam of giant laser
pulses.

The frequency at which sound becomes isothermal may be lower in plasma because of increase in the thermal conductivity [see Eq. (I.23)]. The ion temperature should remain of the
order of 300°K but the electron temperature should reach 10^5°K [81]. However, theoretical estimates [82] show that the equilibrium temperature in a plasma is established in 10^{-10} sec.
Therefore, it would be very interesting to determine the time required for the appearance and
development of the stimulated Brillouin scattering, as compared with the time needed for the
formation of a plasma. One should bear in mind that up to four Stokes and one anti-Stokes component of the Brillouin scattering are observed in oxygen. If we attribute these components to
regeneratively amplified scattering (§8), we find that a long time (~40 nsec) would be needed
for the appearance of all the components. Therefore, it would be interesting to determine the
mechanism responsible for the appearance of the Brillouin components.

The questions raised in the preceding paragraphs were tackled in a special study [53] of
the time dependence of the integrated intensity of the stimulated Brillouin scattered light in
gaseous nitrogen at 150 atm.

TABLE 1

Gas	P, atm	n	$f \cdot 10^{-9}$, Hz	$\Lambda \cdot 10^5$, cm	$\alpha\Lambda$	Our data		v_{ad}, m/sec	v_{is}, m/sec
						$\Delta\nu \cdot 10^2$, cm^{-1}	velocity of hypersound, m/sec		
Nitrogen	125	1.035	0.84	3.3	0.06	2.8 ± 0.1	280 ± 10	386 [80]	297
Oxygen	150	1.038	0.99	3.3	0.06	3.3 ± 0.3	230 ± 30	331	280
Hydrogen	95	1.012	3.3	3.4	0.14	11 ± 1	1130 ± 100	1400	1127
Helium	140	1.005	2.6 (calc.)	3.5 (calc.)	1.7	—	—	1008	783

§10. Time Evolution of the Stimulated Brillouin Scattering in Gaseous Nitrogen at 150 atm

The apparatus described in §7 allowed us to determine simultaneously the time dependence of the integrated intensities of the forward- and back-scattered light (this was done with the aid of FÉK-09 photodiodes) as well as the integrated and time-scanned spectrograms of the scattered light (this was done with the aid of Fabry–Perot interferometers and a PIM-3 image converter).

The overall time resolution was 1.5-2 nsec and it was limited by the time needed for the establishment of the interference pattern in the Fabry–Perot interferometer. The giant laser pulses were of 40-50 MW power and of 20 nsec duration. Light was focused with an $f = 3$ cm lens inside a chamber filled with gaseous nitrogen at 150 atm. The spectrum of the stimulated Brillouin scattering included up to four Stokes components.

Curve 1a in Fig. 16 represents the shape of a giant pulse after transmission under conditions such that neither plasma nor the stimulated Brillouin scattering were generated. Curve 2c corresponds to a pressure of 1 atm at which the scattering did not appear but a plasma was produced. Curve 2d corresponds to a pressure of 150 atm at which the scattering took place and a plasma was generated: in this case the shape of the exciting pulse (curve b) was recorded with a photodiode Fd_1 (the plane-parallel plate m_3 occupied the position denoted by the dashed line in Fig. 11).

A comparison of the curve plotted in Fig. 16 demonstrates that the interaction of the back-scattered radiation with the stimulated radiation in the laser itself altered the shape of the laser pulse and this could delay the moment of formation of a plasma.

Curves 1a, 1b, 2c, and 2d were made to coincide at the points at which a noticeable deviation from the horizontal was first observed.

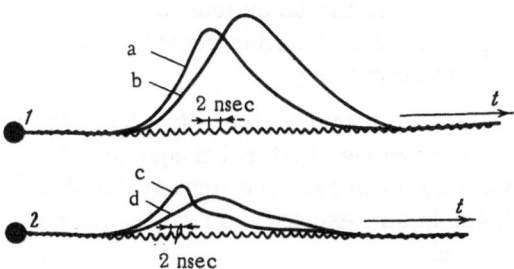

Fig. 16. Combined oscillograms of the radiation incident on and transmitted by a gas-filled chamber. 1) Giant laser pulse: a) in the absence of feedback with the scattering region; b) in the presence of feedback with the scattering region in a high-pressure (150 atm) chamber. 2) Oscillograms of the radiation transmitted by the high-pressure chamber: c), d) nitrogen at 1 and 150 atm, respectively.

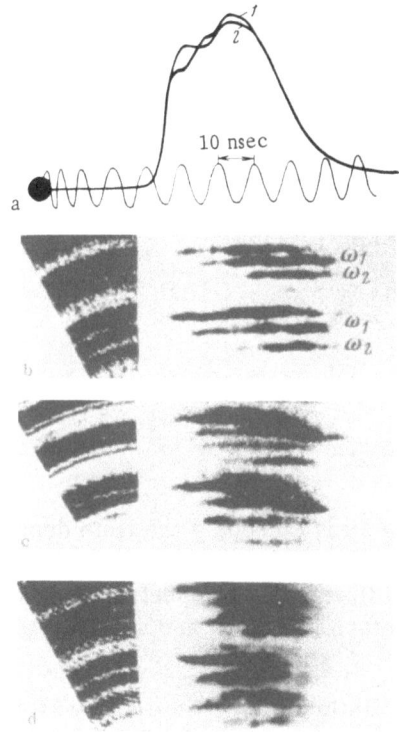

Fig. 17. Simultaneously recorded oscillograms and interferograms of the radiation scattered in nitrogen at 150 atm: a) oscillograms produced by Fd_1; b)-d) Fabry–Perot interferograms produced by F-P_2 (on the left) and the stimulated Brillouin scattering spectra produced by a time-scanned converter (on the right). The time scale is the same in interferograms and oscillograms and the direction of the time arrow is from left to right.

In transmitted light the laser radiation pulsations (curve 2c) were practically insignificant whereas they were clearly observed in the back-scattered radiation (curves 1 and 2 in Fig. 17a). This suppression of the pulsations was likely to be due to the interaction between the back-scattered Brillouin components and the stimulated radiation in the laser itself.

Calorimetric measurements indicated that the total energy of the transmitted light which generated plasma at 1 atm was 40% lower than the energy of the light transmitted at 150 atm when the Brillouin scattering preceded the formation of a plasma.

In regeneratively amplified scattering the Brillouin components appeared at intervals of 7-8 nsec, which was in agreement with the geometrical dimensions of the apparatus. A plasma was generated 8-10 nsec later than at an atmospheric pressure when no Brillouin scattering was observed (curves 2c and 2d in Fig. 16 and curve 1 in Fig. 17a).

The alternative to the successive Brillouin scattering was the mechanism in which a Stokes component of frequency ω_1 reached an intensity sufficient to generate another component of frequency ω_2 without amplification in the ruby crystal. In this case the ω_2 component traveling along the laser beam was recorded interferometrically at the moment of its appearance whereas the ω_1 component was recorded after its amplification in the laser (curve 2 in Fig. 17a, and Figs. 17c and 17d). Such a scattering, which appeared in the interaction region of the light waves, will be called the local multiple stimulated Brillouin scattering.

In some cases the ω_2 component consisted of two lines of different intensity. The stronger line was shifted by ~ 0.003 cm^{-1} and coincided with one of the modes (natural frequencies) of the resonator. This shift could be explained in the same way as in §8. The light at the shifted frequency, scattered through 180° with respect to the laser beam, excited the nearest (in respect of the frequency) axial laser mode because of feedback.

The positions of the components of the stimulated Brillouin scattering were used to calculate the velocity of hypersound in nitrogen: this velocity was found to be 320 ± 20 m/sec, which was less than the adiabatic value.

The results obtained indicate that the stimulated Brillouin scattering can and does appear before the formation of a plasma. After the formation of a plasma the scattering is not observed. Therefore, the isothermal velocity of sound obtained in our measurements and the values reported in [21, 57] cannot be explained by the hypothesis of a rise in the thermal conductivity. On the other hand, it is clear that the velocity of hypersound was determined under conditions in which the stimulated Brillouin scattering occurred in a growing field of the exciting laser radiation (Figs. 16 and 17).

The presence of the anti-Stokes components of the stimulated Brillouin scattering can be explained on the basis of the observation that these components can appear as a result of regenerative amplification in the laser or because of local multiple scattering without such amplification. The anti-Stokes components are observed in gases (this is also true of the results reported in [56]). If the local multiple mechanism is active, strong hypersonic waves may be generated in the forward and the backward direction in the scattering volume and this may give rise to the anti-Stokes component.

The simultaneous occurrence of the local multiple and regeneratively amplified scattering may explain the appearance of a large number of the Brillouin scattering components in gases in spite of the fact that the formation of a plasma suppresses the scattering.

§11. Dependence of the Shifts of the Brillouin Scattering Components on the Intensity of the Exciting Radiation

The question why the velocity of sound deduced from the positions of the components of the stimulated Brillouin scattering is close to the isothermal value has not yet been answered. Since this value was obtained using single-mode radiation when feedback was suppressed [22], it seemed desirable to study at least qualitatively the dependence of the positions of the Brillouin scattering components on the intensity of the exciting radiation because the velocity of hypersound of $\sim 10^9$ Hz frequency determined from the thermal scattering had the adiabatic value [39, §§2, 3].

The apparatus shown schematically in Fig. 12 was used to investigate the stimulated Brillouin scattering in gaseous nitrogen at 125 atm. The laser radiation consisted of a single mode, which was $\Delta\nu \sim (0.5-1) \times 10^{-2}$ cm^{-1} wide. The investigation was carried out in the presence of feedback between the scattering region and the laser (the condition corresponding to the successive appearance of several Brillouin scattering components) and in the absence of such feedback (the scattering region and the laser were isolated by a quarter-wave plate). In the latter case, only one Stokes component of the stimulated Brillouin scattering was observed.

Table 2 lists the measured shifts of the Brillouin components and the velocities of hypersound in compressed nitrogen deduced from these shifts.

The exciting radiation of ~ 100 MW power was focused by lenses of different focal length. The shifts of the Brillouin components were determined using lenses with $f = 30$ and 20 cm. They corresponded to the adiabatic velocity of sound (386 m/sec) at T = 300°K and 125 atm [80].

The average values of the shifts of the components obtained in the regeneratively amplified Brillouin scattering were in agreement, within the experimental error of ± 0.002 cm^{-1}, with the shift of the Stokes component in the absence of feedback. When the focal length was reduced, i.e., when the intensity of the exciting radiation in the interaction region was raised, the velocity of hypersound decreased. The difference between the shifts of the regeneratively amplified

TABLE 2

Focal length of lens f, cm	Shift of Brillouin components, cm^{-1}			Velocity of hypersound v, m/sec
	I*	II		
	$\Delta\nu$	$\Delta\nu$	$\Delta\nu_{av}$	
30	0.038	0.040 0.038	0.039	393±20
20	0.039	0.037 0.037 0.036	0.037	372±20
10	0.035	0.036 0.031 0.033	0.033	331±25
5	0.032	0.028 0.029 0.034	0.030	300±25
3	0.031	0.025 0.027 0.033 0.037	0.030	300±25

* I represents investigations of the stimulated Brillouin scattering in the absence of feedback between the laser and the scattering region (feedback suppressed by a quarter-wave plate); II is the case of successive appearance of several regeneratively amplified Brillouin scattering components.

scattering components obtained with the aid of short-focus lenses was sometimes greater than the experimental error or the separation between the axial laser modes (0.005 cm^{-1}).

Measurements of the shifts of each amplified component under feedback conditions indicated that when long-focus lenses (f = 20-30 cm) were used, the Brillouin components were equidistant and the velocity of sound had the adiabatic value. When the focal length was reduced, the shifts of the amplified components were no longer equidistant: the separation between them decreased and then increased again. The corresponding velocity of sound, deduced from the average value of these shifts, was now less than the adiabatic velocity.

Table 3 lists some of the values of the velocity of hypersound in compressed nitrogen deduced from the positions of the Brillouin components obtained by different authors under different experimental conditions.

It is evident from this table that the velocities of hypersound at 30-150 atm, deduced from the shifts of the Brillouin components, were smaller than the adiabatic values. The stimulated Brillouin scattering used in the measurement of the velocity of hypersound was excited by laser radiation of the highest powers listed in Table 2.

It was reported in [21, 22, 56] that the velocity of hypersound at 159 atm or higher pressures was close to the adiabatic value. When the pressure was lowered, the velocity deviated from its adiabatic value, passed through a minimum at ~50 atm, and rose again reaching the adiabatic value of ~25 atm [21, 57].

Thus, the deviation of velocity of hypersound from the adiabatic value (Tables 2 and 3) was governed not only by the gas pressure but also by the intensity of the laser radiation in the scattering region. When long-focus (f = 20-30 cm) lenses were used, the velocity of hypersound

TABLE 3

Laser performance characteristics*	Focal length of lens f, cm	Gas pressure P, atm	Velocity of hypersound		Reference
			exper. v, m/sec	theor. [80] v, m/sec	
$U = 0.8$ J $\tau = 8 \div 35$ nsec	15.20	25—50	370—290	353—362	[57] †
$U = 0.5$ J $\tau = 6 \div 15$ nsec	5	70—150	337—385	370—399	[21]
$I = 10 \div 100$ MW/cm²	30	125	356±20 ‡	386	[22] †
$I = 250$ MW/cm²	3	125	280±10	386	[20, § 9]
$I = 50$ MW/cm²	3	150	320±20	399	[53, § 10]
$I = 100$ MW/cm²	4	250	435±20	450±5	[56]
		500	560±25	590±5	

* I is the intensity, U is the energy, and τ is the duration of a laser pulse.

† The stimulated Brillouin scattering observed in the absence of feedback between the laser and the scattering region (feedback suppressed by a quarter-wave plate).

‡ Results of measurements of the velocity of hypersound, refined in [80].

was adiabatic ([22] and Table 2). When the focal length of the lenses employed was reduced, the velocity of hypersound decreased ([20, 53] and Table 2).

Similar results were obtained also for other gases [20, 21, 57] (see also Table 1).

The discrepancy between the experimental results reported in [20-22] (Table 3) was attributed in [80] to the presence of several modes in the exciting radiation because this would give rise to large errors in the determination of the shifts of the Brillouin scattering components. However, this was not true of our results because they were obtained using single-mode laser radiation in the absence of any feedback between the scattering medium and the laser. It should be mentioned that the results obtained by different workers were compared ignoring the inhomogeneous distribution of the radiation across the laser beam.

We mentioned earlier (§1) that because of the absorption of sound the position of the maximum (Ω_{max}) in the thermal Brillouin scattering does not agree with the frequency (Ω_0) of the elastic thermal wave: according to Eq. (I.6), $\Omega_{max} \to 0$ when the absorption of sound increases.

However, the condition $\alpha \Lambda < \pi$ is satisfied in the studies of the stimulated Brillouin scattering (Table 1). Therefore, in the linear approximation, we may take $\Omega_{max} = \Omega_0$ as being accurate to within 1%. On the other hand, when this condition is satisfied a high-intensity hypersonic wave may be generated in the scattering region. It is known [83] that the absorption coefficient of sound for large-amplitude waves is different from the corresponding coefficient for small-amplitude waves (this is associated with the viscosity and the thermal conductivity of the medium). Since the profile of a large-amplitude acoustic wave changes during its propagation (this change can be regarded as the appearance and growth of high-frequency harmonics) and the absorption coefficient of gases is $\sim \Omega^2$, it is clear that a nonlinear distortion of the wave profile should be accompanied by an increase in the absorption of sound. The absorption coefficient of a large-amplitude acoustic wave is a function of its spectral composition, which can vary during propagation. It follows that the absorption coefficient will also vary in space. Therefore, in contrast to the absorption of small-amplitude waves, for which the absorption coefficient α_0 is constant, the absorption coefficient of large-amplitude waves is a function of the coordinates.

In the case of gases [83, 84], we find that $\alpha = \alpha_0$ Re $\sim (10^4-10^5) \alpha_0$, where Re = $p/\eta_{eff} \Omega$ is the Reynolds number (p is the pressure in the acoustic wave and η_{eff} is the effective viscosity).

Direct measurements of the velocity of large-amplitude ultrasonic waves, carried out with a precision to within the order of the Mach number ($M = p/P_0$) indicated [83] that weak discontinuities travel at the velocity of small-amplitude waves. This allows us to assume that, in contrast to the absorption coefficient whose value is very sensitive to the amplitude of sound, direct measurements of the velocity can be carried out even when the acoustic waves have large amplitudes.* In the case of stimulated Brillouin scattering in compressed gases, the Mach number is $M \approx 0.1$.

However, it is difficult to see a priori how an increase in the absorption of sound can affect the shift of the components of the stimulated Brillouin scattering. The influence of the absorption can be found by solving the system of nonlinear Maxwell and gas-dynamic equations, but this presents difficulties. It is possible that the experimentally observed dependence of the shift of the Brillouin components on the amplitude of sound is due to the increase in the absorption resulting from the nonlinearity of the large-amplitude acoustic waves.

The reduction in the velocity of hypersound can also be explained qualitatively in other ways. Since the deviations from the adiabatic value are usually observed when the appearance of a plasma at the focus of a lens suppresses the stimulated Brillouin scattering, i.e., when the field of the exciting laser radiation grows in intensity, it is possible that the reduction in the velocity of hypersound can be explained as suggested by Askar'yan [85]. He considered the change in the interaction between atoms in a strong light field. The increase in the interaction between atoms in a growing field may give rise to short-lived molecular complexes, which would reduce the velocity of hypersound. This may possibly explain why the deviation of the velocity of hypersound from its adiabatic value is a function of the intensity of the exciting radiation and of the gas pressure.

If molecular complexes with an average lifetime τ appear in a gas as a result of enhancement of the molecular interaction in a laser beam, the velocity of sound will vary with the gas pressure if the period of the hypersonic vibrations is of the same order as τ. Since the formation of complexes can be regarded as resulting from a bimolecular reaction, it follows that a number of complexes should be a function of the gas pressure. Therefore, the velocity of hypersound may — in principle — decrease to $v_{ad}/\sqrt{2}$.

The qualitative considerations given above are not in conflict with the thermodynamic theory. The collision time, i.e., the time during which colliding molecules interact (the lifetime of molecular complexes), is governed by the strength of the attractive forces acting between the molecules. The reduction in the velocity of hypersound with increasing gas density [21] can be related, as indicated by the formula [86]

$$v^2 = v^2_{ad}\left(\frac{V^2}{(V-b)^2} - \frac{2a}{VM}\right),$$

to the action of the attractive forces as represented by an increase in the van der Waals correction a.

The suggested quantitative explanations of the observed variation of the velocity of hypersound, excited in the stimulated Brillouin scattering at high laser radiation intensities, do not provide a final solution of the problems associated with this variation. Further theoretical and experimental investigations would be needed before we could understand the contributions of different processes affecting the shifts of the Brillouin scattering components at high exciting radiation intensities.

---

* It should be pointed out that the velocity of a strong shock wave ($M > 1$, $Re \gg 1$) relative to an unperturbed medium is greater than the velocity of sound [83].

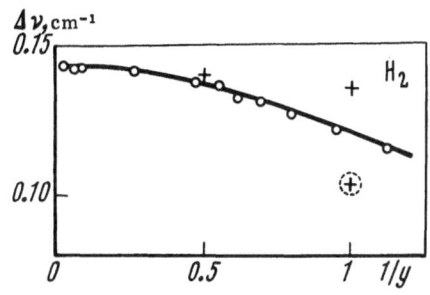

Fig. 18. Dependence of the shift of the Stokes
component of the stimulated Brillouin scatter-
ing on $1/y = \pi^{3/2}\bar{l}/\Lambda$ (open circles). The plus
sign represents the position of the maximum of
the spectral intensity of the fine structure in the
thermal scattering, considered in the adiabatic
approximation [32]; the plus sign in a circle
represents the same maximum but in the isothermal approximation (see Figs. 1 and 2).

§12. Stimulated Brillouin Scattering in Low-Pressure Hydrogen

The results of our measurements (§11) indicated that the adiabatic velocity of hypersound was obtained when long-focus lenses were used, i.e., when the intensities of the exciting radiation in the interaction region were relatively low, irrespective of the presence of any feedback between the scattering region and the laser. At these intensities the light scattered in low-pressure gases would have to be amplified before it could be recorded photographically (§5).

Under feedback conditions the additional amplification of the Stokes component of the Brillouin radiation scattered at $\theta = 180°$ could be obtained as a result of superregenerative gain in a laser mode (§8).

The present author used trigger amplification of the Brillouin components in a laser in a spectroscopic investigation of the stimulated Brillouin scattering in gaseous hydrogen at 2-100 atm. Two equidistant Stokes components of the Brillouin scattering were observed at high pressures (70-100 atm), but only one Stokes component appeared at low pressures (2-70 atm).

Figure 18 shows the dependence of the shift $\Delta\nu$ of the Brillouin components on the ratio $\bar{l}/\Lambda$ or on the quantity $1/y$ $(y = \pi^{-1/2} \Lambda/\bar{l})$, where $\bar{l}$ is the mean free path of the gas molecules and Λ is the wavelength of hypersound (§3). The experimental points plotted in Fig. 18 were obtained by averaging the results deduced from several batches of photographs in which the positions of the Brillouin components were found to vary smoothly with the gas pressure. The smoothness of the variation and the precision of the results were governed by the separation between the axial modes in the laser resonator ($\Delta\nu = c/2L \approx 5 \times 10^{-3}$ cm^{-1}).

The dependence of the absorption coefficient of sound on the ratio of the frequency of sound f to the pressure p in hydrogen [87] suggested that the absorption was entirely due to the shear viscosity beginning from $f/p \approx 10$ MHz/atm. According to the relaxation theory [15] the velocity of hypersound should then be close to v_∞.

In view of the aforementioned results, it was possible to describe the propagation of hypersonic waves in our experiments ($f/p \geq 50$ MHz/atm) by the gas-kinetic Boltzmann equation for an ideal gas (§3), since the compression energy was not converted (in the time available) into the energy of the rotational degrees of freedom of the molecules.

It is evident from Fig. 18 that the shift of the Brillouin components at 10-100 atm was 0.141-0.138 cm^{-1}. The velocity of sound, calculated from this shift, 1480-1486 ± 60 m/sec, was in agreement with the adiabatic value $v_\infty = 1400$ m/sec.

When the pressure was lowered ($1/y \geq 0.5$) the shift of the Brillouin components decreased more rapidly than predicted by the adiabatic theory. In this range of pressures the mean free path of molecules became comparable (in order of magnitude) with the wavelength of hypersound. Consequently, the reduction in the shift of the Brillouin components in hydrogen in the $1/y > 0.5$ range could be attributed to a gradual transition from the adiabatic to the isothermal propagation of sound.

CHAPTER IV

EXPERIMENTAL RESULTS OF SPECTROSCOPIC
INVESTIGATIONS OF THE NONLINEAR ELECTRO-
CALORIC INTERACTION OF HIGH-POWER LIGHT
PULSES WITH A GAS

§13. Investigations of the Spectra of the Amplified Stimulated Temperature Scattering in Low-Pressure Hydrogen

The theory of the stimulated temperature scattering (§5), which is valid in the hydrodynamic approximation [58], predicts that the anti-Stokes component should decay and the Stokes component should grow in accordance with the law $E_1^2(x) = E_1^2(0) \times \exp\{g_{1T}x\}$. The spectrum of the stimulated temperature scattering differs considerably from the thermal scattering spectrum. Instead of a monotonic fall of the intensity on both sides of the exciting line, which occurs in the thermal scattering resulting from fluctuations in the entropy, the stimulated scattering consists of only the Stokes component and, most importantly, has a maximum at a frequency which is shifted by $\Omega_m = \chi |K - K_1|^2$ from the frequency of the exciting line. In other words, if the width of the unshifted line in the thermal scattering is given by Eq. (I.11), $\delta\omega_c = 2\chi |K - K_1|^2$, the lowest threshold value of the exciting radiation power corresponds to the half-width of the thermal scattering line, i.e., to the frequency $\Omega_m = \frac{1}{2}\delta\omega_c$.

The present author investigated gaseous hydrogen because the width of its scattering line was considerably greater than the corresponding widths observed for other gases [1].

It was shown in §5 that in order to observe the stimulated temperature scattering it would be necessary to amplify the scattered light quite strongly. In the present investigation the amplification of the back-scattered light ($\theta = 180°$) was provided by the superregenerative gain in that laser mode whose frequency was located at the gain maximum [Eq. (I.35)].

The most effective amplification of the stimulated temperature scattering in the laser resonator was achieved by using a mode separator which ensured that the stimulated Brillouin scattering was amplified much less strongly.

Figure 19 shows the spectrograms of the amplified stimulated temperature scattering recorded at various pressures. The dependence of the position of the stimulated temperature scattering line (or the half-width of this line) on the quantity $1/y = \pi^{3/2}\bar{l}/\Lambda$ is plotted in Fig. 20. It is evident from this figure that in the $y \geq 2$ range the experimental values of the half-width of the undisplaced fine-structure line were in agreement with the calculated values. In this case the line profile was Lorentzian and the width was proportional to $1/y$.

A deviation from the linear dependence was observed in the $y < 2$ range. When the value of y was reduced, the shift of the maximum of the stimulated temperature scattering approached $\sim 0.11-0.12$ cm^{-1}. This was in qualitative agreement with the conclusions of the gas-kinetic theory [32, 36] for an ideal gas with a Maxwellian distribution of the molecular velocities (§3): in compressed gases, when $\bar{l}/\Lambda \ll 1$, the central line was accompanied by the Brillouin components; in the $\bar{l}/\Lambda > 1$ case, there were no Brillouin components and only the central line was observed. The central line was Lorentzian at high pressures ($\bar{l}/\Lambda \ll 1$) whereas the profile of the thermal scattering line approached the Doppler shape at low pressures when $\bar{l}/\Lambda \gg 1$ or $1/y \to \infty$. The shift of the stimulated temperature scattering component, equal to the half-width of the undisplaced line in the thermal scattering, ranged from a small value at high pressures to the

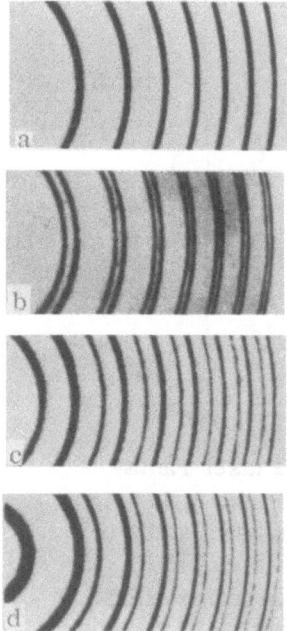

Fig. 19. Spectrograms of the amplified stimulated temperature scattering in gaseous hydrogen at various pressures: a) laser radiation; b) P = 6.5 atm; c) P = 2.5 atm; d) P = 1.5 atm. Spectral range of the Fabry–Perot interferometer 0.166 cm^{-1}.

Fig. 20. Dependence of the position of the stimulated temperature scattering line (Ω_m) on 1/y. The linear dependence is predicted by Eq. (I.38) in the hydrodynamic approximation.

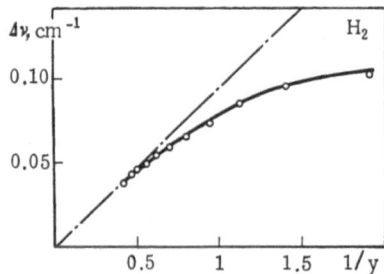

Doppler-half-width limit at low pressures (for hydrogen at room temperature this limit was 0.12 cm^{-1}, as shown in Fig. 1).

§14. Broadening of the Frequency Spectrum of High-Power Laser Pulses in Gases

We observed an anti-Stokes broadening of the laser radiation spectrum in nitrogen and hyd hydrogen at 1-150 atm. This broadening was studied using the apparatus shown schematically in Fig. 12. The feedback between the laser and the scattering region was suppressed by a quarter-wave plate.

A light pulse of ~ 150 MW power was focused with an f = 3 cm lens in a gas-filled chamber. The observations were carried out in the forward and backward directions with respect to the laser beam. The light scattered through 180° was studied in the same way as the stimulated Brillouin scattering. The spectral range of the spectrometer was 0.166, 0.5, or 1.66 cm^{-1}.

The anti-Stokes broadening of the frequency spectrum was observed in the forward but not in the backward direction. Only the Stokes component of the Brillouin scattering (§11) was found in the light scattered through 180° in nitrogen at 50-150 atm.

Figure 21 shows microphotograms of the spectra of the laser radiation transmitted by

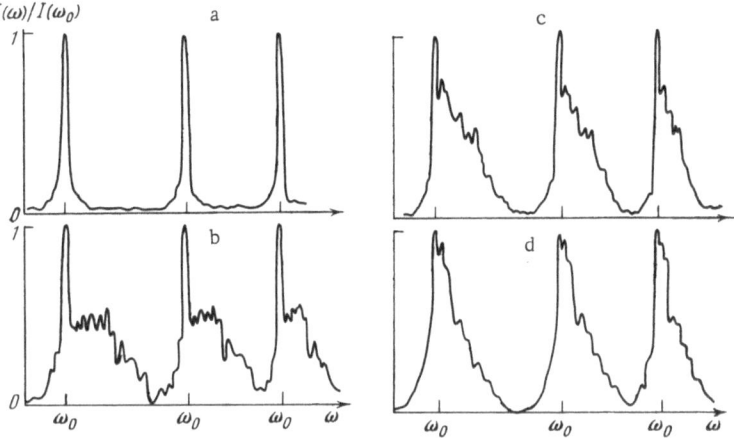

Fig. 21. Microphotograms of the spectra of the laser radiation transmitted across the focusing region, I(ω)/I(ω_0), in
gaseous nitrogen at various pressures: a) unfocused laser
radiation; b) spectrogram of light transmitted across the
focusing region at 15 atm; c), d) spectra of light transmitted
by the focusing region at 35 and 100 atm, respectively. Spectral range of the Fabry–Perot interferometer $\Delta\nu_{FP} = 0.6$
cm^{-1}.

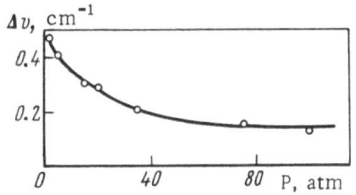

Fig. 22. Pressure dependence of the average
half-width of the spectrum of the radiation
transmitted by the focusing region.

the scattering region in gaseous nitrogen kept at various pressures. It is evident from this
figure that the anti-Stokes broadening did not form a monotonic background but had some
structure.

The intensity of the maxima rose with decreasing frequency shift and with increasing
pressure. When the pressure was lowered, the general broadening of the spectrum increased.

Figure 22 shows the dependence of the average half-width of the spectrum on the pressure
in nitrogen. It is evident that the broadening increased approximately threefold when the pressure was lowered from 100 to 1 atm. When the intensity of the laser radiation was halved, the
broadening decreased twofold. The broadening was most probably due to the time dependence
of the nonlinear part of the refractive index of the gas in the laser radiation field which varied
with time (this gave rise to variation of the phase of the light wave with time).

Several recent theoretical [88-90] and experimental [91, 92] investigations indicated that
a manifold broadening of the frequency spectrum of optical pulses of 10^{-8}-10^{-9} sec duration
may occur in self-focusing channels in liquids or in other nonlinear media. Such broadening
may occur without a significant distortion of the amplitude envelope of the initial pulse [88].
Let us now consider whether such broadening can explain the results obtained in the present
study. We shall use the first-approximation formulas describing the propagation of a quasi-
harmonic plane wave in a nonlinear medium [88].

If the largest phase shift Φ_{max} exceeds π, it follows from [88] that the width of the spectrum of a pulse should be close to the maximum frequency deviation

$$\Delta\omega = \frac{\partial\Phi}{\partial t} = -\frac{Kx}{n_0}\left(\frac{\partial n'}{\partial t}\right)_{max}, \qquad (IV.1)$$

where $n = n_0 + n'(E^2)$ is the refractive index and n' is the nonlinear part of this index.

The "critical" distance x_{cr} in which Φ_{max} becomes of the order of π is

$$x_{cr} = \pi n_0/Kn'_{max}. \qquad (IV.2)$$

The profile $I(\omega)$ of the spectral line of a radiation pulse transmitted by a nonlinear medium depends on changes in the pulse intensity with time. For example, in the case of trapezoidal and parabolic profiles we may expect a maximum separated by $\Delta\omega$ from the undisplaced laser line; here, $\Delta\omega$ is given by Eq. (IV.1) in the Stokes and anti-Stokes regions.* The processes of broadening in the direction of the Stokes and anti-Stokes regions are separated in time. During the rise of the pulse intensity $(\partial n'/\partial t) < 0$ and, therefore, $\Delta\omega > 0$ whereas during the fall of this intensity $(\partial n'/\partial t) > 0$ and, therefore, $\Delta\omega < 0$.

Under the experimental conditions in our study a plasma was produced in the focusing region and this plasma was opaque to the laser beam [92]. Under these conditions breakdown occurred when the intensity of the laser radiation increased sufficiently (§10). Therefore, the broadening of the spectrum could be recorded only during the rise of the laser pulse intensity.

We shall explain the anti-Stokes broadening of the spectrum during the rise of the laser pulse intensity by assuming that the nonlinear part of the refractive index changes due to the electrocaloric effect [13, 58]. Then,

$$n' \approx \varepsilon'/2n_0 = \tfrac{1}{2}n_0\left(\frac{\partial\varepsilon}{\partial T}\right)_p\Delta T, \qquad \Delta T = \frac{T\left(\frac{\partial\varepsilon}{\partial T}\right)_p E^2}{8\pi\rho C_p}. \qquad (IV.3)$$

In this case $(\partial n'/\partial t) < 0$ and $\Delta\omega > 0$.

We shall assume that no self-focusing takes place and that the volume of the focusing region remains constant (this volume is determined by the lens parameters). We shall then estimate the intensity of light necessary for the observation of the effect in question.

If the focusing region is 0.1 cm long and x_{cr} is at least an order of magnitude smaller (0.01 cm), it follows from Eqs. (IV.2) and (IV.3) that the spectral broadening in nitrogen at 1-100 atm should be observed in electric fields of the light wave amounting to $\sim 10^7$-10^8 V/cm. Before breakdown and the consequent formation of an opaque plasma the local fields in the focusing region can be $\sim 5 \times 10^7$ V/cm [78].

Under these conditions the broadening predicted by Eq. (IV.1) should be 0.1 cm^{-1}, which is in order-of-magnitude agreement with the observed broadening (Fig. 22).

The structure of the spectral broadening (Fig. 21) can be explained by the space–time evolution of a laser pulse: stimulated emission spreads transversely (from the center to the periphery of the ruby laser) in a time comparable with the pulse duration: the pulses generated by different parts of the end face of the ruby can have different shapes and they can be shorter than the giant pulse emitted by the whole end face [93].

---

*Equation (IV.1) ignores the fine structure over frequency intervals of the order of the initial width of the laser radiation spectrum.

It follows that the rise of the intensity with time will be different in different parts of the focal region. Since the spectrograms record the broadening integrated over the whole focal region and the profile of the broadened line depends, in accordance with Eq. (IV.1), on the time dependence of the intensity at each point in the focal region, we observe just one maximum.

At low gas pressures the breakdown occurs after a longer delay than at high pressures [78]. Consequently, the broadening of the laser radiation spectrum may be stronger at low pressures since at these pressures the derivative $\partial n'/\partial t$ is higher than at high pressures, as confirmed qualitatively by the experimental results shown in Figs. 21 and 22.

The author is deeply grateful to I. L. Fabelinskii and D. I. Mash for suggesting the subject, directing the work, and their constant interest, to A. N. Oraevskii and V. S. Starunov for their valuable advice, and to V. P. Zaitsev, M. V. Vetokhin, M. A. Vysotskaya, V. V. Korobkin, V. A. Morozov, V. S. Starunov, and M. Ya. Shchelev for their great help in the experimental investigations.

LITERATURE CITED

1. I. L. Fabelinskii, Molecular Scattering of Light, Plenum Press, New York (1968).
2. L. I. Mandel'shtam, Zh. Russ. Fiz.-Khim. Obshchst., Chast Fiz., 58:381 (1926).
3. L. Brillouin, Ann. Phys. (Paris), 17:88 (1922).
4. G. S. Landsberg, Selected Works [in Russian], Izd. AN SSSR, Moscow (1958).
5. E. Gross, Z. Phys., 63:685 (1930).
6. E. Gross, Nature, 126:201, 400, 603 (1930).
7. I. L. Fabelinskii, Experimental and Theoretical Investigations in Physics (G. S. Landsberg Memorial Collection) [in Russian], Izd. AN SSSR, Moscow (1959).
8. M. A. Leontovich, Z. Phys., 72:247 (1931).
9. M. A. Leontovich, Izv. Akad. Nauk SSSR, Ser. Fiz., No. 5, p. 633 (1936).
10. I. L. Fabelinskii, Usp. Fiz. Nauk, 63:355 (1957); 77:649 (1962).
11. V. L. Ginzburg, Dokl. Akad. Nauk SSSR, 42:168 (1944).
12. L. D. Landau and E. M. Lifshitz, Electrodynamics of Continuous Media, Pergamon Press, Oxford (1960).
13. H. Lamb, Hydrodynamics, 6th ed., Cambridge University Press (1932).
14. H. O. Kneser, Ann. Phys. (Leipzig), 16:337 (1933).
15. L. I. Mandel'shtam and M. A. Leontovich, Zh. Eksp. Teor. Fiz., 7:438 (1937).
16. C. S. Venkateswaran, Proc. Indian Acad. Sci., A15:316 (1942).
17. G. S. Landsberg, Usp. Fiz. Nauk, 36:284 (1948).
18. R. Lennuier, C. R. Acad. Sci., 226:708 (1948).
19. J. Cabannes, C. R. Acad. Sci., 226:710 (1948).
20. D. I. Mash, V. V. Morozov, V. S. Starunov, and I. L. Fabelinskii, ZhETF Pis. Red., 2:562 (1965).
21. D. H. Rank, T. A. Wiggins, R. V. Wick, D. P. Eastman, and A. H. Guenther, J. Opt. Soc. Amer., 56:174 (1966).
22. E. E. Hagenlocker and W. G. Rado, Appl. Phys. Lett., 7:236 (1965).
23. D. P. Eastman, T. A. Wiggins, and D. H. Rank, Appl. Opt., 5:879 (1966).
24. I. L. Fabelinskii, D. I. Mash, V. V. Morozov, and V. S. Starunov, Phys. Lett., 27A:253 (1968).
25. Inelastic Scattering of Neutrons in Solids and Liquids (Proc. Second Intern. Symp., Vienna, 1960), International Atomic Energy Agency, Vienna (1961).
26. L. Van Hove, Phys. Rev., 95:249 (1954).
27. M. Nelkin and S. Yip, Phys. Fluids, 9:380 (1966).

28. R. Kubo et al., Statistical Mechanics, Interscience, New York (1965).

29. L. P. Kadanoff and P. C. Martin, Ann. Phys. (N.Y.), 24:419 (1963).

30. F. V. Hunt, in: American Institute of Physics Handbook (ed. by D. E. Gray), 1st ed., McGraw-Hill, New York (1957).

31. J. M. J. Van Leeuwen and S. Yip. Phys. Rev., 139:A1138 (1965).

32. M. Nelkin and A. Ghatak, Phys. Rev., 135:A4 (1964).

33. S. Yip and M. Nelkin, Phys. Rev., 135:A1241 (1964).

34. A. G. Gibbs and J. H. Ferziger, Phys. Rev., 138:A701 (1965).

35. S. Yip and S. Ranganathan, Phys. Fluids, 8:1956 (1965).

36. S. Ranganathan and S. Yip, Phys. Fluids, 9:372 (1966).

37. P. L. Bhatnagar, E. P. Gross, and M. Krook, Phys. Rev., 94:511 (1954).

38. M. Nelkin and S. Yip, Phys. Fluids, 9:380 (1966).

39. T. J. Greytak and G. B. Benedek, Phys. Rev. Lett., 17:179 (1966).

40. R. Y. Chiao, C. H. Townes, and B. P. Stoicheff, Phys. Rev. Lett., 12:592 (1964).

41. R. G. Brewer and K. E. Rieckhoff, Phys. Rev. Lett., 13:334a (1964).

42. R. G. Brewer, Appl. Phys. Lett., 5:127 (1964).

43. D. I. Mash, V. V. Morozov, V. S. Starunov, E. V. Tiganov, and I. L. Fabelinskii, ZhETF Pis. Red., 2:246 (1965).

44. E. Garmire and C. H. Townes, Appl. Phys. Lett., 5:84 (1964).

45. N. M. Kroll, Bull. Amer. Phys. Soc., 9:222 (1964); J. Appl. Phys., 36:34 (1965).

46. C. L. Tang, J. Appl. Phys., 37:2945 (1966).

47. N. Bloembergen, Nonlinear Optics, Benjamin, New York (1965).

48. E. Garmire, F. Pandarese, and C. H. Townes, Phys. Rev. Lett., 11:160 (1963).

49. N. M. Kroll, Phys. Rev., 127:1207 (1962).

50. Y. R. Shen and N. Bloembergen, Phys. Rev., 137:A1787 (1965).

51. E. E. Hagenlocker, R. W. Minck, and W. G. Rado, Phys. Rev., 154:226 (1967).

52. A. N. Oraevskii, Molecular Oscillators [in Russian], Nauka, Moscow (1965).

53. V. V. Korobkin, D. I. Mash, V. V. Morozov, I. L. Fabelinskii, and M. Ya. Shchelev, ZhETF Pis. Red., 5:372 (1967).

54. Proc. Intern. Conf. on Physics of Quantum Electronics, San Juan, Puerto Rico, 1965, publ. by McGraw-Hill, New York (1966).

55. R. W. Minck, R. W. Terhune, and C. C. Wang, Appl. Opt., 5:1595 (1966).

56. S. Dumartin, B. Oksengorn, and B. Vodar, C. R. Acad. Sci. B, 262:1680 (1966).

57. T. A. Wiggins, R. V. Wick, and D. H. Rank, Appl. Opt., 5:1069 (1966).

58. V. S. Starunov, Phys. Lett., 26A:428 (1968).

59. G. I. Zaitsev, Yu. I. Kyzylasov, V. S. Starunov, and I. L. Fabelinskii, ZhETF Pis. Red., 6:802 (1967).

60. V. S. Starunov, Dissertation [in Russian], Physics Institute, USSR Academy of Sciences, Moscow (1965).

61. Yu. I. Kyzylasov and V. S. Starunov, ZhETF Pis. Red., 7:160 (1968).

62. V. V. Morozov, Prib. Tekh. Eksp., No. 2, p. 179 (1968).

63. B. S. Guberman and V. V. Morozov, Opt. Spektrosk., 22: 673 (1967).

64. D. I. Mash, V. V. Morozov, V. S. Starunov, and I. L. Fabelinskii, ZhETF Pis. Red., 2:41 (1965).

65. S. V. Krivokhizha, D. I. Mash, V. V. Morozov, V. S. Starunov, and I. L. Fabelinskii, ZhETF Pis. Red., 3:378 (1966).

66. B. L. Borovich, V. S. Zuev, and V. A. Shcheglov, Zh. Eksp. Teor. Fiz., 49:1031 (1965).

67. M. Hercher, Appl. Phys. Lett., 7:39 (1965).

68. B. H. Soffer and B. B. McFarland, Appl. Phys. Lett., 8:166 (1966).

69. D. Röss, Z. Naturforsch., 20a:696 (1965).

70. C. R. Giuliano and L. D. Hess, Appl. Phys. Lett., 9:196 (1966).

71. N. G. Basov, A. Z. Grasyuk, I. G. Zubarev, and L. V. Tevelev, Tr. Fiz. Inst. Akad. Nauk SSSR, 31:74 (1965).
72. V. N. Smiley, Appl. Opt., 5:977 (1966).
73. A. A. Vuylsteke, J. Appl. Phys., 34:1615 (1963).
74. N. G. Basov, V. N. Morozov, and A. N. Oraevskii, Zh. Eksp. Teor. Fiz., 49:895 (1965).
75. W. R. Sooy, Appl. Phys. Lett., 7:36 (1965).
76. S. Tolansky, High Resolution Spectroscopy, Methuen, London (1947).
77. E. K. Zavoisky and S. D. Fanchenko, Appl. Opt., 4:1155 (1965).
78. Yu. P. Raizer, Usp. Fiz. Nauk, 87:29 (1965).
79. R. G. Brewer, Appl. Phys. Lett., 9:51 (1966).
80. W. M. Madigosky, A. Monkevicz, and T. A. Litovitz, J. Acoust. Soc. Amer., 41:1308 (1967).
81. L. Spitzer, Jr., Physics of Fully Ionized Gases, 2nd ed., Wiley, New York (1962).
82. S. L. Mandel'shtam, P. P. Pashinin, A. M. Prokhorov, Yu. P. Raizer, and N. K. Sukhodrev, Zh. Eksp. Teor. Fiz., 49:127 (1965).
83. L. K. Zarembo and V. A. Krasil'nikov, Introduction to Nonlinear Acoustics [in Russian], Nauka, Moscow (1966).
84. D. I. Mash, V. V. Morozov, V. S. Starunov, and I. L. Fabelinskii, Zh. Eksp. Teor. Fiz., 55:2053 (1968).
85. G. A. Askar'yan, ZhETF Pis. Red., 6:672 (1967).
86. I. G. Mikhailov, V. A. Solov'ev, and Yu. P. Syrnikov, Fundamentals of Molecular Acoustics [in Russian], Nauka, Moscow (1964).
87. J. J. Markham, R. T. Beyer, and R. B. Lindsay, Rev. Mod. Phys., 23:353 (1961).
88. L. A. Ostrovskii, ZhETF Pis. Red., 6:807 (1967).
89. F. DeMartini, C. H. Townes, T. K. Gustafson, and P. L. Kelley, Phys. Rev., 164:312 (1967).
90. R. G. Brewer, Phys. Rev. Lett., 19:8 (1967).
91. K. Shimoda, Jap. J. Appl. Phys., 5:615 (1966).
92. S. L. Mandelstam, P. P. Pashinin, A. M. Prokhorov, and N. K. Sukhodrev, Proc. Intern. Conf. on Physics of Quantum Electronics, San Juan, Puerto Rico, 1965, publ. by McGraw-Hill, New York (1966), p. 548.
93. R. V. Ambartsumyan, N. G. Basov, V. S. Zuev, P. G. Kryukov, V. S. Letokhov, and O. B. Shatberashvili, Zh. Eksp. Teor. Fiz., 51:406 (1966).

EXPERIMENTAL INVESTIGATIONS OF THE THERMAL AND STIMULATED MOLECULAR SCATTERING OF LIGHT IN SOLUTIONS IN A WIDE RANGE OF WAVELENGTHS *

I. M. Aref'ev

Investigations of binary liquid solutions were carried out by light scattering methods. The work was concentrated on three principal topics: 1) measurements of the width of the central component of the scattering spectrum of solutions and determination of diffusion coefficients; 2) measurements of the velocity of hypersound at about 5 GHz and of the dispersion of this velocity in solutions far from and close to the critical solution point; 3) stimulated scattering of light in solutions. The investigations were carried out over the whole spectrum of scattered light, in greater or lesser detail. The spectroscopic apparatus used had a resolving power $\sim 10^3$ to $\sim 4 \times 10^{10}$. The investigated solutions were: acetone–carbon disulfide; bromoform–normal propyl alcohol; tertiary butyl alcohol–water; triethylamine–water; carbon disulfide–carbon tetrachloride.

INTRODUCTION

Light is scattered in pure liquids because of deviations of the refractive index from its average value. If the intensity of the incident light is such that its interaction with the scattering medium is negligibly weak, light is scattered by thermal fluctuations of the density, anisotropy, and polarizability of molecules. Fluctuations of the density and anisotropy of molecules are responsible for what is known as the Rayleigh scattering of light, with fluctuations of the density giving rise to a polarized Rayleigh line and fluctuations of the anisotropy to a wide depolarized wing of the Rayleigh line [1]. Fluctuations of the polarizability of molecules, whose frequencies are of the order of natural molecular vibrations, give rise to the Raman scattering of light. Fluctuations of the density in pure liquids are due to entropy or temperature fluctuations, which disperse at a rate governed by the thermal diffusivity, and pressure or volume fluctuations, which travel at the velocity of sound. Therefore, the Rayleigh line of a liquid is a triplet consisting of an undisplaced central component and two Brillouin satellites.† The separation between the Brillouin components is governed by the frequency of the thermal wave responsible for the scattering.

---

* Condensed text of a thesis submitted for the degree of Candidate of Physicomathematical Sciences, defended on December 23, 1968 at the P. N. Lebedev Physics Institute, Academy of Sciences of the USSR. Scientific supervisor: I. L. Fabelinskii.

† Brillouin scattering is usually referred in the Soviet literature as the Mandel'shtam–Brillouin scattering.

The distribution of the intensity in the wing of a Rayleigh line is quite complex. The near part of the wing (up to ~ 10 cm^{-1}) is due to modulation of the incident radiation by rotational jumps of molecules from one equilibrium position to another (rotational diffusion of molecules); the far part of the wing (up to ~ 100-150 cm^{-1}) is due to modulation of the incident radiation by rocking oscillations (librations) of molecules which occurs in the time between two jumps [1-3]. Starunov, Tiganov, and Fabelinskii [4-6] and later Stegeman and Stoicheff [7] found a fine structure in the near part of the Rayleigh line of some viscous liquids. This fine structure was due to the scattering by thermal shear waves predicted theoretically by Leontovich [8] and Rytov [9].

In the case of solutions the spectrum of the thermal scattering of light includes also a narrow line at the central (undisplaced) frequency, which is due to the scattering of light by fluctuations of the concentration which disperse at the rate of interdiffusion of molecules of the components forming the solution. Fluctuations of the pressure, entropy, and concentration in solutions are independent of one another and are random variables [1]. Therefore, the spectra due to the scattering by optical inhomogeneities resulting from these fluctuations can be discussed independently. The general nature of the spectrum of the thermal scattering of light in a solution, corresponding to a nonzero scattering angle θ, is shown schematically in Fig. 1.

Fluctuations of the concentration vary with time and space in accordance with the diffusion equation. The spectrum of light scattered by optical inhomogeneities resulting from these fluctuations is given by a dispersion curve with a half-width $\delta\omega_c$ given by [1]

$$\delta\omega_c = 8n^2 Dk^2 \sin^2 \frac{\theta}{2}. \tag{1}$$

Here, n is the refractive index of the solution; D is the diffusion coefficient; k is the wave vector of the incident light. The spectrum of light scattered by optical inhomogeneities resulting from fluctuations of the entropy is also described by a dispersion curve with a half-width given by Eq. (1) except that the diffusion coefficient D must be replaced with the thermal diffusivity χ. The scattering of light by optical inhomogeneities due to fluctuations of the entropy and concentration will be referred to simply as the entropy and concentration scattering, respectively. The relevant orders of magnitude are $\chi \sim 10^{-3}$ cm^2/sec and $D \sim 10^{-5}$ cm^2/sec. This means that the concentration scattering line should be approximately 100 times narrower than the entropy scattering line. For equal integrated scattering intensities the spectral density in the concentration

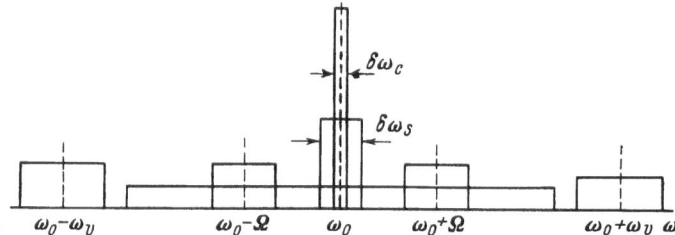

Fig. 1. Spectrum of the thermal scattering of light in a solution (schematic representation): $\delta\omega_c = 8n^2 Dk^2 \cdot \sin^2(\theta/2)$, $\delta\omega_s = 8n^2\chi k^2 \sin^2(\theta/2)$, $\Omega = 2n\omega_0 v/c \sin(\theta/2)$, where D is the diffusion coefficient; χ is the thermal diffusivity; k is the wave vector of the incident light; θ is the scattering angle; n is the refractive index of the solution; v is the velocity of sound; c is the velocity of light; Ω is the frequency of the acoustic wave; ω_v is the frequency of natural vibrations of molecules.

scattering will be approximately 100 times higher than the spectral density of the entropy scattering. Thus, in solutions with a significant concentration scattering the width of the central component is governed by the diffusion coefficient, which can be found by measuring this width.

The question now arises how to measure the width of the central component. It follows from Eq. (1) that for n ~ 1.5, incident radiation of $\lambda = 632\text{Å}$, and $\theta = 90°$, the half-width is $\delta\nu_c = \delta\omega_c/2 \sim 60$ kHz or $\delta\nu_c \sim 4 \times 10^{-6}$ cm^{-1}. Measurements of spectral intervals of this order can be made if the spectroscopic apparatus has a resolving power of $\sim 10^{10}$. If we use a Fabry–Perot interferometer, which is frequently employed in studies of the fine structure of the Rayleigh line, we find that the separation between mirrors of R = 0.95 reflectivity, would have to be ~ 50 m and one would have to maintain constant temperature and pressure in this gap to at least 10^{-4} deg and 10^{-4} mm Hg, respectively. Clearly, it would be hopeless to attempt measurements by this method. The problem can be solved by optical heterodyning in which extremely high resolving powers (up to 10^{14}) can be achieved. This method was first suggested by Gorelik [10] and put into practice by Forrester, Gudmundsen, and Johnson [11]. The physical basis of the optical heterodyne methods is given in Chap. I.

One of the aims of the present investigation was to master the optical heterodyne methods and to carry out measurements of the widths of the central components in the scattering spectra of solutions exhibiting considerable concentration scattering. Apparatus for square-law detection of light with a resolving power of $\sim 4 \times 10^{10}$ was assembled. Measurements were made of the widths of the central components of the scattering spectra and the diffusion coefficients were deduced for solutions of acetone in carbon disulfide and of bromoform in n-propanol under normal conditions, i.e., at room temperature and atmospheric pressure. The results, which had not been obtained before, are presented in Chap. II.

In recent years the optical heterodyning of light has been used successfully in measurements of the widths of the central components in the spectra of light scattered by the suspensions of macromolecules [12-13], binary solutions near the critical solution points [14-16], and pure liquids near their critical points [17-18] and under normal conditions [19]. The cited investigations represent all the work that has been published on the molecular scattering by the heterodyne methods. This work is reviewed in Chap. I.

The relatively high intensity of the concentration scattering makes it difficult to observe the fine structure of the Rayleigh line in solutions if the excitation source is a mercury-discharge lamp which has a wide emission line.

The first attempt to determine the fine structure of the Rayleigh lines of solutions was made by Sunanda Bai [20]. He found a weak structure in just two cases and he used it to determine the velocity of hypersound. Sterin [21] was able to resolve more clearly the triplet of benzene–toluene mixtures. More systematic investigations of the fine structure of the Rayleigh lines of solutions were carried out by Lanshina and Shakhparonov [22] with the aim of obtaining data on the velocity of hypersound. However, some of the values given in [22] were later found to be too low. The investigations reported in [20-22] were the only ones carried out in the prelaser stage of the study of the Rayleigh triplets of solutions. The development of gas lasers with extremely narrow emission lines made it possible to determine the positions of the Brillouin components in the Rayleigh triplet and to use their shifts in the measurements of the velocity of hypersound in solutions with the same precision as in the case of pure liquids [23-25].

The present author investigated the Brillouin scattering spectra, determined velocity of hypersound, and measured the dispersion of this velocity in aqueous solutions of tertiary butyl alcohol. An analysis of the acoustic data of Burton on the basis of the Mandel'shtam–Leontovich relaxation theory of the propagation of sound [26] indicated that a considerable dispersion of the velocity of sound can be expected for aqueous solutions of tertiary butyl alochol. Such a dispersion was indeed observed: it amounted to 7% for a solution containing 11 mol.% of the alcohol. In pure water the velocity of sound showed no dispersion [6] whereas in pure tertiary butanol it was 2.7 ± 0.9%. The results obtained were used to calculate the critical relaxation frequencies of the bulk viscosity which showed that the relaxation propagation of sound in water–butanol solutions could be studied by direct ultrasonic methods. Investigations of aqueous solutions of tertiary butyl alcohol are described in Chap. III.

We have mentioned already that strong concentration scattering makes it difficult to record the fine structure of the Rayleigh lines of solutions (even under conditions far from critical) if the scattering is excited by a broad mercury line. These difficulties increase greatly on approach to the critical solution point at which phase separation (unmixing) occurs, because at this point the intensity of the concentration scattering rises strongly. This explains why the fine structure of the Rayleigh lines of solutions has not been determined (until very recently) near the critical solution point. Shoroshev, Shakhparonov, and Lanshina [25] recorded the fine structure of the Rayleigh line for isooctane–nitroethane solutions at just one temperature which was relatively far (2.5 deg) from the critical solution point.

The present author recorded for the first time the fine structure of the Rayleigh line of a solution near its critical point. This was done using a neon–helium laser and a suitable Fabry–Perot interferometer. The solution in question was the classical water–triethylamine system with a lower solution point. Later Chen and Polonsky [28] determined the fine structure of the Rayleigh line near the upper solution point of an n-hexane–nitrobenzene system.

The present author used the results obtained to determine the velocity of hypersound and the dispersion of this velocity near the critical solution point. It was found that the temperature dependence of the velocity of hypersound was unusual for a critical point: the velocity increased on approach to this point. An attempt is made to explain the results obtained within the framework of the simple relaxation theory of the propagation of sound [1]. This is done in Chap. IV.

The properties of a scattering medium change in the field of high-power laser radiation. Large fluctuations of the refractive index, which increase in time, are generated. The scattering of the laser radiation by these fluctuations generates a strong stimulated scattering spectrum. In principle, every type of thermal scattering may serve as the basis for a corresponding stimulated effect. This is due to the fact that the field of a light wave may alter all those physical parameters whose thermal fluctuations are responsible for the spectrum shown in Fig. 1. For example, a change in the polarizability of the molecules in the field of a light wave gives rise to the stimulated Raman scattering [29]. Electrostriction produces the stimulated Brillouin scattering [30]. The high-frequency Kerr effect gives rise to the stimulated scattering in the wing of the Rayleigh line [31]. The electrocaloric effect produces the stimulated temperature (entropy) scattering [32].

The concentration scattering in solutions also has its stimulated analog. However, the stimulated concentration scattering has not yet been observed because laser radiation of the necessary power is available only in the form of very short pulses of about 10^{-8} sec duration. the characteristic time of changes in the concentration fluctuations is 10^{-4}-10^{-5} sec. Thus, the concentration instability cannot develop during a laser pulse. However, all the other types of stimulated scattering observed in pure liquids should also occur in solutions.

In the present study the stimulated Brillouin scattering was observed in water–tertiary butanol solutions. This investigation was carried out in order to obtain information on the velocity of hypersound. Since the stimulated scattering spectra could be recorded in a time of $\sim 10^{-8}$ sec, it seemed logical to derive the velocity of hypersound from the stimulated spectra rather than from the thermal spectra because in the latter case the exposures ranged from several minutes to several hours. However, the velocities of hypersound deduced from the stimulated scattering spectra differed from the values obtained from the thermal spectra. The cause of this discrepancy is discussed in Chap. III.

The last chapter (V) is devoted to the stimulated scattering effects in carbon disulfide and in solutions of this compound in carbon tetrachloride. Carbon disulfide is a liquid in which self-focusing of light occurs very easily (see, for example, [33]). The high densities of the laser radiation in the self-focusing filaments ensure the appearance of all the types of stimulated scattering that are observed in liquids. A study of the various scattering spectra of carbon disulfide and of solutions of this compound in carbon tetrachloride reveal amplification of weak R_2-line ruby laser radiation. This effect is attributed to the four-photon interaction in a nonlinear medium which has been treated theoretically by Starunov [3] and Chiao, Kelley, and Garmire [34].

Only recently it was thought that no important new effect could be discovered in the classical scattering of light. The appearance of gas and pulsed solid lasers has transformed the situation so that the scattering of light is now one of the rapidly developing branches of physics. The use of lasers and of various measurement methods such as heterodyning and interferometry made it possible to deduce new important information from the scattering spectra.

CHAPTER I

OPTICAL HETERODYNE METHODS AND SOME OF THEIR APPLICATIONS IN INVESTIGATIONS OF MOLECULAR SCATTERING OF LIGHT

§1. Optical Heterodyne Methods

Heterodyning of light consists of mixing of different optical frequencies in a suitable nonlinear element and of isolation of a signal corresponding to the difference frequency. This mixing effect is achieved if the response of the radiation detector is nonlinear in respect of the amplitude of the light field reaching it. For example, the photomultiplier current i is proportional to the square of the amplitude of the light field E incident on the photocathode:

$$i \propto E^2. \tag{I.1}$$

Let us assume that the incident light E consists of two monochromatic waves:

$$\left. \begin{array}{l} E_1 = E_{01} \cos (\omega_1 t - \varphi_1), \\ E_2 = E_{02} \cos (\omega_2 t - \varphi_2). \end{array} \right\} \tag{I.2}$$

Then, substituting Eq. (I.2) into Eq. (I.1) and performing several simple trigonometric transformations, we obtain

$$i \propto \frac{1}{2} E_{01}^2 + \frac{1}{2} E_{02}^2 + \frac{1}{2} E_{01}^2 \cos (2\omega_1 t - 2\varphi_1) + \frac{1}{2} E_{02}^2 \cos (2\omega_2 t - 2\varphi_2) + \\ + E_{01}E_{02} \cos [(\omega_1 + \omega_2) t - (\varphi_1 + \varphi_2)] + E_{01}E_{02} \cos [(\omega_1 - \omega_2) t - (\varphi_1 - \varphi_2)]. \tag{I.3}$$

Thus, the photocurrent has a constant component and four alternating components of frequencies $2\omega_1$, $2\omega_2$, $\omega_1 + \omega_2$, and $\omega_1 - \omega_2$. The first three frequencies are called the "sum" frequencies and the last is known as the "difference" frequency. In fact, the sum frequencies are absent in the photocurrent spectrum: the inertia of the system is such that it does not respond to these very high frequencies. It follows that the difference-frequency component of the current can be isolated by passing the output signal through a suitable RC filter.

The wavefronts of the beams reaching the photocathode should be collinear. If the difference between the phases $\varphi_1 - \varphi_2$ of the incident light waves varies from zero to 2π over the surface of the photocathode, the net photocurrent will not have the component with the difference frequency. This can be demonstrated quite easily by integrating the last term in Eq. (I.3).

Square-Law Detection

Let us assume that the radiation reaching the photocathode has a certain intensity distribution over the spectrum. To be specific we shall consider the light scattered by fluctuations in the refractive index of the scattering medium. The photocurrent i at any moment t is

$$i(t) \propto E^2(t) = (\delta E(t) e^{-i\omega_0 t})^2,$$

where ω_0 is the frequency of the light incident on the scattering medium. The fluctuations $\delta E(t)$ are proportional to fluctuations of the permittivity of the scattering medium. Since the photocurrent does not contain components with the optical frequencies, it follows that $i(t) \propto |E(t)|^2$.

We shall now introduce a function which describes the correlation in fluctuations of the light field $R_1(\tau)$ and the correlation function of fluctuations of the current $R_2(\tau)$:

$$\left. \begin{array}{l} R_1(\tau) = \overline{E^*(t) E(t+\tau)}, \\ R_2(\tau) = \overline{E^2(t) E^2(t+\tau)}. \end{array} \right\} \tag{I.4}$$

We now find that

$$R_2(\tau) = \overline{E^2(t) E^2(t+\tau)} = \overline{[E(t) E(t+\tau)]^2} = \overline{E^*(t) E(t+\tau)}^2 = R_1^2(\tau). \tag{I.5}$$

According to the well-known Wiener–Khinchin theorem [17] the correlation functions of fluctuations are related to their spectral intensities by the expressions

$$\left. \begin{array}{l} I(\omega) = \int\limits_{-\infty}^{\infty} R_1(\tau) e^{i\omega\tau} \, d\tau, \\ i(\omega) = \int\limits_{-\infty}^{\infty} R_2(\tau) e^{i\omega\tau} \, d\tau. \end{array} \right\} \tag{I.6}$$

The integrals are normalized so that, for example, $\int\limits_{-\infty}^{\infty} I(\omega) \, d\omega = \dfrac{1}{2\pi R_1(0)}$. We then have

$$R_1(\tau) = \int\limits_{-\infty}^{\infty} I(\omega) e^{-i\omega\tau} \, d\omega,$$

$$i(\omega) = \int\limits_{-\infty}^{\infty} R_1^2(\tau) e^{i\omega\tau} \, d\tau = \int\limits_{-\infty}^{\infty} R_1(\tau) e^{i\omega\tau} \int\limits_{-\infty}^{\infty} I(\omega') e^{-i\omega'\tau} \, d\omega' \, d\tau =$$

$$= \int\limits_{-\infty}^{\infty}\!\!\int R_1(\tau) e^{i(\omega-\omega')\tau} \, d\tau \, I(\omega') \, d\omega' = \int\limits_{-\infty}^{\infty} I(\omega - \omega') I(\omega') \, d\omega'. \tag{I.7}$$

It follows that

$$i(\omega) = \int\limits_{-\infty}^{\infty} I(\omega - \omega') I(\omega') \, d\omega'.$$

The photocurrent spectrum is the convolution of the spectrum of the radiation incident on the photocathode with itself. The integrals used in Eq. (I.7) are well known. The convolution of a dispersion curve with itself is also a dispersion curve but with doubled half-width. The convolution of a Gaussian curve with itself is also a Gaussian curve but with a half-width 1.414 times larger than the initial value, etc. (see, for example, the review [35]).

It follows that when the radiation incident on the photocathode has a spectrum which can be described by a dispersion curve with a half-width $\delta\omega$, the spectrum of the photocurrent is also a dispersion curve with a half-width $2\delta\omega$. The spectrum of the photocurrent has a peak at the zero frequency and contains formally all the frequencies from zero to infinity, whereas actually the range of frequencies is from zero to the highest transmission frequency of the detector. Thus, an analysis of the photocurrent spectrum yields the distribution of intensity in the incident radiation spectrum.

We shall follow Forrester [36] and estimate the signal-to-noise ratio S/N for square-law detection. We shall assume that the noise spectrum of the detector is dominated by the shot component: $N = 2ei_0 \Delta f$ is the Schottky formula, where e is the electronic charge, i_0 is the total detector current, and Δf is the band width.

According to Forrester [36], we obtain

$$\left(\frac{S}{N}\right)_{\omega=0} = \frac{i_0}{e\delta'},\tag{I.8}$$

where the current i_0 corresponds to the total power of the radiation incident on the detector, $i_0 = \int\limits_0^\infty I(\omega) \, d\omega$, and δ' is the half-width of the incident radiation expressed in hertz.

Let us consider a photocathode receiving radiation from a helium–neon laser ($\lambda = 6328\text{Å}$) of $P = 10^{-2}\,\text{W}$ power and 10^3 Hz half-width. If we assume that the photon–electron conversion coefficient is $\beta = 0.01$, we find that the current is

$$i_0 = \frac{P\beta e}{h\nu} \approx \frac{10^{-2} \cdot 10^7 \cdot 10^{-2} \cdot 5 \cdot 10^{-10} \cdot 6,3 \cdot 10^{-5}}{6,6 \cdot 10^{-27} \cdot 3 \cdot 10^{10}} \approx 10^5 \text{ cgs esu.}$$

Then,

$$\left(\frac{S}{N}\right)_{\omega=0} \approx \frac{10^5}{5 \cdot 10^{-10} \cdot 10^3} = 2 \cdot 10^{11}.\tag{I.9}$$

This very high value of the signal-to-noise ratio indicates that it is extremely easy to achieve square-law detection of the gas-laser radiation. In practice, this method is employed to measure the widths of the emission lines of cw gas lasers [37].

In the case of square-law detection of the scattered laser radiation the signal-to-noise ratio is considerably smaller because the molecular scattering coefficients are small. For example, the 90° scattering coefficient for a liquid under normal conditions is $R_{90} = 10^{-6}$ cm^{-1} for $\lambda = 6328\text{Å}$ [1]. By definition, the scattering coefficient is

$$R = \frac{I}{I_0} \frac{L^2}{V},\tag{I.10}$$

where I_0 is the intensity of the incident light; I is the intensity of the scattered light; V is the volume of the scattering region; L is the distance from the scattering volume to the point of observation. Thus, the signal-to-noise ratio in square-law detection of the scattered light can be found by multiplying the right-hand of Eq. (I.8) by RV/L^2.

In heterodyning of the scattered light it is necessary to place an aperture-restricting stop of ~ 1 mm diameter in front of the photocathode. This is necessary for two reasons: 1) if the aperture is large, the beat signal is averaged over the surface of the photocathode and the total signal decreases; 2) the scattered radiation must be recorded at a definite angle. Therefore, V ~ 0.01 cm³. If we assume that L ~ 10 cm and consider solutions for which the scattering coefficient is of the same order as the coefficient for liquids ($R_{sol} \sim R_{liq}$), we find that the signal-to-noise ratio for $\delta' \sim 10^4$ Hz (Chap. II) is

$$\left(\frac{S}{N}\right)_{\omega=0} \sim 2.$$

Thus, under the conditions given above, the signal-to-noise ratio in square-law detection of the light scattered by solutions under normal conditions is approximately unity. For pure liquids this ratio is two orders of magnitude smaller. We can therefore see that square-law detection of the scattered light is not as easy as the corresponding detection of laser radiation.

Optical Heterodyning

In the optical heterodyne methods a photocathode receives not only the radiation spectrum $I(\omega)$ being investigated but also a monochromatic beam of frequency ω_0. The photocurrent spectrum then represents a convolution of the spectrum being investigated and of the delta function $\delta(\omega')$:

$$i(\omega) = \int_{-\infty}^{\infty} I(\omega' - \omega)\,\delta(\omega')\,d\omega', \tag{I.11}$$

where

$$\delta(\omega') = I_0 \quad \text{for} \quad \omega' = \omega_0,$$
$$\delta(\omega') = 0 \quad \text{for} \quad \omega' \neq \omega_0.$$

We thus find that

$$i(\omega) = I_0\, I(\omega_0 - \omega). \tag{I.12}$$

Thus, under optical heterodyning conditions the photocurrent spectrum reflects exactly the spectrum of the radiation being investigated. The source of monochromatic radiation (the heterodyne) is naturally a laser. Since in all cases (with the possible exception of the critical points) we have

$$I_0 \gg I(\omega)\,\delta\omega, \tag{I.13}$$

it follows that the shot noise of a detector is entirely due to the heterodyne signal and the signal-to-noise ratio is

$$\frac{S}{N} = \frac{2\pi}{e}\, I(\omega_0 - \omega). \tag{I.14}$$

If we multiply and divide the right-hand side of Eq. (I.14) by the half-width of the spectrum being investigated, $\delta\omega$, we obtain the same expression for the signal-to-noise ratio as in square-law detection.

Since under optical heterodyning conditions the detector noise is entirely due to the heterodyne signal, we can modulate the radiation being investigated and "extract" the useful signal from the noise by a suitable modulation and synchronous detection system [36]. Let us now summarize the main advantages and disadvantages of the square-law detection and optical heterodyning methods.

Advantages of optical heterodyning: 1) the spectrum of the current at the output is an exact image of the spectrum of the incident radiation; 2) the sensitivity is high if modulation and synchronous detection systems are used.

Disadvantages of optical heterodyning: 1) it is necessary to achieve exact coincidence of the wave fronts of the radiation being investigated and of the heterodyne signal; 2) the detection system is complex.

Advantages of square-law detection: 1) only the radiation being investigated has to be detected; 2) the detection system is relatively simple.

Disadvantages of square-law detection: 1) it is necessary to convert the photocurrent spectrum to the incident radiation spectrum; 2) the sensitivity of the detection system is less than in the case of optical heterodyning.

The basic principle of the optical heterode methods is clear but considerable difficulties are encountered in practical applications. These difficulties arise because of the need for very sensitive and complex apparatus. This is why these methods have not become standard in investigations of the scattering of light.

§2. Experimental Investigations of the Thermal Scattering of Light in Liquids and Solutions by Optical Heterodyne Methods

After the possibility of heterodyning of light was established experimentally [11], the first investigations of the molecular scattering of light were carried out by this method. These investigations started with materials having high scattering coefficients. These were pure liquids and solutions near their critical points. However, there are also some materials which have large scattering coefficients under normal conditions: they are suspensions and solutions of macromolecules. The suspended particles execute Brownian motion. This motion modulates the incident light and the scattering spectrum has a line at the undisplaced frequency. The width of this line is given by Eq. (1), where D is the linear diffusion coefficient. According to the Stokes–Einstein relationship [38], this diffusion coefficient is given by

$$D = \frac{kT}{6\pi\eta r}, \qquad (I.15)$$

where k is the Boltzmann constant; T is the absolute temperature; η is the viscosity of the solvent; r is the radius of a particle or a macromolecule. Equation (I.15) is, strictly speaking, applicable only to spherical microparticles in dilute solutions.

The diffusion coefficient of macromolecules is $D \sim 10^{-8}$-10^{-7} cm^2/sec [13]. Hence, the spectrum of light scattered at an angle of 90° should have a line with the half-width $\delta\nu \sim 10$-100 Hz. Cummins, Knable, and Yeh [12] used a helium–neon laser ($\lambda = 6328$Å) to build apparatus for the optical heterodyning of light (a homodyne spectrometer) with an instrumental half-width of 6 Hz, i.e., with a resolving power of $\sim 10^{14}$. They measured the half-width of the central component in the scattering spectra of dilute solutions of monodisperse polystyrene mole- cules. The spectra were recorded for different scattering angles. Cummins et al. used their results in combination with Eqs. (1) and (I.15) to determine the radii of these molecules. Later Dubin, Lunacek, and Benedek [13] used a spectrometer with square-law detection and a resolv- ing power of $\sim 10^{14}$ to measure the width of the central component of the scattering spectra and to determine the diffusion coefficients for aqueous solutions of natural and synthetic macro- molecules (dioxyribonucleic acid, some proteins, polystyrene, etc.). The values of the diffusion coefficients obtained in this way were in good agreement with the values deduced by other meth- ods. Dubin et al. confirmed also that the angular dependence of the half-width of the scattering line obeyed closely Eq. (1).

The investigations of Cummins et al. and of Dubin et al. [12, 13] demonstrated that hetero- dyning of light could be used to obtain important information on biological and chemical macro- molecules.

The intensity of the scattering I_c resulting from fluctuations of the concentration in binary molecular solutions is given by the formula [1]:

$$I_c = I_0 \frac{V^* \pi^2 V}{2\lambda^4 L^2} \left(\frac{\partial \varepsilon}{\partial c} \right)^2_{\rho, T} \overline{\Delta c^2} (1 + \cos^2 \theta), \tag{I.16}$$

$$\overline{\Delta c^2} = ckT \left/ \frac{\partial p}{\partial c} V^*, \right. \tag{I.17}$$

where ε is the permittivity of the solution; c is the concentration; ρ is the density; V^* is the volume of a single fluctuation; the rest of the notation has been explained earlier.

On approach to the critical solution temperature the rate of change of the osmotic pres- sure with the concentration $\partial p/\partial c$ tends to vanish and fluctuations of the concentration $\overline{\Delta c^2}$ of Eq. (I.17) and, consequently, the intensity of the scattered light of Eq. (I.16) both increase with- out limit. The diffusion coefficient D is related to fluctuations of the concentration by the ex- pression [39]

$$D = bTc(1 - c)^2 / N\overline{\Delta c^2}, \tag{I.18}$$

where b is the mobility of a solute molecule in a dilute solution; N is the total number of parti- cles in the volume being considered; T is the temperature in ergs. Consequently, if fluctuations of the concentration become stronger, the diffusion coefficient must decrease. This reduction in the diffusion coefficient close to the critical solution point was confirmed experimentally by physicochemical methods in several papers of Krichevskii et al. [42]. At the critical point it- self the diffusion coefficient should vanish.

It follows from the kinetics of fluctuations of the concentration that the intensity of the central component in the scattering spectrum should increase on approach to the critical point and the width of this component should decrease (at the critical point itself it should vanish). The profile of the central component, which describes the time dependence of the correlation function of fluctuations of the concentration, should remain of the dispersion type.

The first measurements of the profile of the central component near the critical solution point were carried out by Alpert, Yeh, and Lipworth [14, 41] who used a homodyne spectrometer [12]. They investigated a solution consisting of 53 wt.% of cyclohexane and 47 wt.% of aniline, whose critical solution temperature was $t_C = 29.69°C$. They established that between 30.05 and 29.69°C the width of the scattering line decreased linearly at a rate of ~ 200 Hz/deg. The line profile remained of the dispersion type. These measurements were carried out down to $(t - t_C)$ ~ 0.01 deg. Approximation to the critical point gave the value of the residual width of the scat- tering line, which was ~ 40 Hz for $\theta = 20.5°$. Alpert et al. did not decide whether this residual width was instrumental, resulting from the fact that the assumed t_C was not equal to the true t_C, or a new physical result which did not agree with theoretical predictions. Subsequent measurements of Chu [16] and of White, Osmundson, et al. [15], carried out on other solutions, also failed to resolve this point. Chu determined the half-width of the central component of the light scattered in an aqueous solution of isobutyric acid for $(t - t_C) = 0.36$ deg and different scattering angles. The half-width corresponding to $\theta = 60°$ was $\nu = 230$ Hz.

At large scattering angles the dependence (1) was not obeyed and the results of measurements were described by the empirical formula

$$\frac{\Delta \nu}{2} = 6.32 \cdot 10^4 \, K^2 \, (1 - 4.5) \cdot 10^4 \, K^2, \text{ where } K = \frac{4\pi n \sin \frac{\theta}{2}}{\lambda}.$$

White et al. [15] determined the width of the central component of the light scattered in a cyclo-hexane—polystyrene solution near the critical point. They confirmed the angular dependence of Eq. (1) but they found that the half-width varied sublinearly with $(t - t_C)$. This result was in conflict with those reported in [14, 41].

Measurements of the scattering spectra provide the only method for investigating diffusion in the immediate vicinity of the critical solution point because the condition of constant temperature within very narrow limits means that measurements must be carried out rapidly, which can be done by optical heterodyning but not by physical chemistry methods. An interesting problem, which has not yet been resolved, was formulated by Leontovich [39]. According to his conclusions the reduction in the diffusion coefficient near the critical point of vaporization of a solution [42] does not mean that particles move very slowly but that one particle is immediately replaced by another. This hypothesis on the kinetics of fluctuations of particles in solutions near the critical point can be checked by investigations of the scattering spectra.

According to the theory given in [1, 43], near the critical point the central component of the scattering spectrum of a pure liquid behaves exactly as the central component of the scattering spectrum of a solution. The intensity of light scattered by isobaric fluctuations of the density (or the entropy) is given by the formula*

$$I_S = I_0 V^* \frac{\pi^2 V}{2\lambda^4 L^2} \left(\rho \, \frac{\partial \varepsilon}{\partial \rho} \right)_T^2 \left(\frac{1}{\rho} \, \frac{\partial \rho}{\partial S} \right)_p^2 \overline{\Delta S^2} \, (1 + \cos^2 \theta), \tag{I.19}$$

where fluctuations of the entropy are given by

$$\overline{\Delta S^2} = k C_p \rho / V^*. \tag{I.20}$$

The specific heat at constant pressure C_p becomes infinite at the critical point [44] and the scattering intensity increases.

We have mentioned earlier that in those cases when fluctuations of the entropy can be described by the diffusion equation, the line corresponding to the scattering by these fluctuations has the dispersion profile and the half-width

$$\delta \omega_S = 8n^2 \chi k^2 \sin^2 \frac{\theta}{2}. \tag{I.21}$$

Since the thermal diffusivity χ is

$$\chi = \frac{\varkappa}{C_p \rho}, \tag{I.22}$$

---

*Formulas (I.19) and (I.16) are not valid in the immediate vicinity of the critical point [1].

where χ is the thermal conductivity, it follows that in absence of singularities in χ and ρ the tendency for C_p to become infinite implied that the scattering line becomes narrower and, in the limit, its width vanishes.

The first and so far the only experimental investigations of the undisplaced frequency in the scattering spectra obtained in the critical region are the studies of Ford and Benedek [17, 45] and of Alpert et al. [18].

Ford and Benedek [17] investigated sulfur hexafluoride SF_6 whereas Alpert et al. [18] studied carbon dioxide. The measurements were carried out using a spectrometer with square-law detection [17] or a homodyne spectrometer with a resolving power of $\sim 10^{14}$ [18]. Both investigations confirmed the dispersion profile of the scattering spectrum and the linear dependence of the half-width on $\sin^2 \theta/2$. According to Ford and Benedek [17], the half-width of the scattering spectrum $\Delta\nu$ was a linear function of $(T - T_C)$ and the half-width was finite at $(T - T_C) = 0$ because measurements were carried out along an isochore, which was displaced slightly away from the critical isochore. Alpert et al. [18] also reported that $\Delta\nu$ was a linear function of $(T - T_C)$ and they found that the residual width at the critical point was ~ 100 Hz for $\theta = 10°$. Alpert et al. concluded that this residual width was governed by the molar volume V and they gave the following theoretical formula for the half-width of the central scattering line near the critical solution point:

$$\Delta\nu = \frac{16\pi}{\lambda^2}\frac{\varkappa}{\rho_0}\frac{\sin^2\dfrac{\theta}{2}}{RT}\left[(T - T_C) + \frac{3}{4}T_C\left(\frac{V - V_C}{V_C}\right)^2\right]. \tag{I.23}$$

Here, R is the gas constant. Alpert et al. [18] intended to investigate in greater detail Eq. (I.23).

An estimate of the half-width of the central component in the scattering spectrum of a pure liquid far from the critical point gives $\nu_S \sim 10^7$ Hz or 3×10^{-4} cm^{-1} for $\theta = 90°$. It is practically impossible to determine profiles of this half-width with a Fabry–Perot interferometer. The optical heterodyne methods are more promising but even then it is very difficult to deal with such narrow lines because of the small molecular scattering coefficients of liquids. Nevertheless, Lastovka and Benedek [19] were able to overcome the difficulties and to measure the central component in the scattering spectrum of toluene, which exhibits strong molecular scattering. The scattering was investigated in the angular range $0.33 \leq \theta \leq 3°$. At these low values of the scattering angle the spectral density of the scattered radiation is higher because the integrated scattering intensity of Eq. (I.19) increases with decreasing θ whereas the half-width of Eq. (I.21) decreases. Lastovka and Benedek used an optical heterodyne method. The heterodyne source was a laser beam scattered by irregularities in the window of a cuvette placed inside the laser resonator (this heterodyne method has an important disadvantage in failing to provide means for controlling the intensity of the heterodyne beam). The resolving power of the apparatus ranged from 5×10^{11} to 2×10^{14}, depending on the spectral interval. The threshold sensitivity of the system was $\sim 10^{-14}$ W. In the range of the scattering angles investigated by Lastovka and Benedek the spectrum had an exact dispersion profile with a half-width ranging from 75 to 7500 Hz. This half-width was a linear function of K^2 [$K = 2nk \sin(\theta/2)$]. The slope of this linear dependence yielded the thermal diffusivity $\chi = (0.88 \pm 0.02) \times 10^{-4}$ cm^2/sec. The diffusivity calculated from thermodynamic parameters had the value $\chi_{stat} = (0.97 \pm 0.05) \times 10^{-4}$ cm^2/sec.

The results obtained by Lastovka and Benedek indicated that the theory given in [1] was applicable to the liquid being investigated, i.e., the thermal diffusivity equation described exactly the time and space dependences of fluctuations of the entropy for fluctuation wavelengths in the range $\lambda_f \approx (0.1-2) \times 10^{-4}$ mm. Moreover, these results confirmed that the theory given in [1] could be used to interpret the spectrum of light scattered by liquids near the gas—liquid transition point and they showed that the scattering spectrum provides a sensitive and precise method for measuring the thermal diffusivity without limitation as to wavelength, as is the case in conventional thermodynamic measurements ($\lambda_f = \infty$). Consequently, it should be possible to use the scattering method in investigations of liquids whose thermal diffusivity is a function of the wavelength or frequency of thermal fluctuations.

§3. Heterodyning of Light under Stimulated Scattering Conditions

Light can be heterodyned also under stimulated scattering conditions. In this case a beam of light being heterodyned is allowed to fall on a fast-response photodiode whose signal is applied to a fast-response oscillograph. The beat signal is a sinusoid which modulates the laser pulse. The separation between maxima or minima of the sinusoid is equal to the period of the difference oscillation T and its reciprocal 1/T gives the frequency ν in hertz. The range of frequencies which can be investigated in this way is limited. The upper limit is set by the time resolution of the oscillograph, which is typically $\sim 10^{-9}$ sec (for I2-7 oscillographs). The lower limit is determined by the duration of the Q-switched laser pulses which is $\sim 10^{-8}$ sec. Thus, we can study spectra falling in the frequency range 10^8-10^9 Hz, i.e., 100 MHz to 1 GHz. Obviously, only an optical heterodyne method can be used.

An optical heterodyne method has been used in two investigations of the stimulated scattering of light [46, 47] in which measurements were made of the Brillouin components. The frequency of the optical beats $\nu_L - \nu_B$ (ν_L is the laser frequency and ν_B is the frequency of a Brillouin component) of most liquids lies in the range 4-7 GHz if the scattering angle is $\theta = 180°$. Therefore, the Brillouin scattering spectrum should be studied using small angles θ. Jennings and Takuma [46] observed stimulated Brillouin scattering in carbon disulfide placed in a nonaxial resonator (the axis of this resonator made an angle of 0.0433 rad with the direction of the ruby laser beam) [48]. The stimulated Brillouin radiation was mixed with the laser radiation in a fast-response photodiode. The beat frequency was 123 MHz. The formula

$$\Delta\omega = 2\omega_0 \frac{v}{c} n \sin\frac{\theta}{2} \tag{I.24}$$

was used to determine the velocity of hypersound v. This velocity was v = 1223 ± 24 m/sec, which was in agreement (within the limits of the experimental error) with the value deduced from the thermal scattering spectra [1].

Brewer [47] measured the frequency shift of the Brillouin components in the scattering of light in benzene, toluene, and acetone. The scattering angle was 180° and the frequency shifts were 6.34, 5.70, and 4.49 GHz, respectively. These beat frequencies were measured by mixing a photodiode signal of frequency $\nu_L - \nu_B$ in a nonlinear detector with the radiation produced by a klystron, so that the frequency of the final beat signal was convenient to measure.

Optical heterodyning of the Brillouin scattered radiation provides a convenient method for measuring the dispersion of the velocity of sound. However, the values of the velocity of hypersound obtained from the stimulated scattering spectra differ from the values deduced from the thermal scattering spectra and the cause of this discrepancy is not yet fully understood. This point will be discussed in detail in Chap. III.

CHAPTER II

MEASUREMENT OF THE WIDTH OF THE CENTRAL
COMPONENT OF THE THERMAL SCATTERING
SPECTRUM AND DETERMINATION OF INTERDIFFUSION
COEFFICIENTS OF SOLUTIONS OF ACETONE IN CARBON
DISULFIDE AND OF BROMOFORM IN n-PROPANOL

§1. Spectrometer with Square-Law Detection

In the present investigation the width of the central component in the scattering spectra of solutions was measured by the square-law detection method. The decisive factor in the selection of the measurement method was its simplicity. A spectrometer [49] was assembled as shown schematically in Fig. 2. An LG-35 neon–helium laser was used as the excitation source. The laser emitted $\lambda = 6328\text{Å}$ radiation and its output power was about 15 mW (multimode emission). This particular laser was chosen because it was possible to restrict the radiation to the axial TEM_{n00} modes by reducing the aperture of a diaphragm within the resonator. When this was done the laser radiation consisted of equidistant longitudinal modes separated by $\Delta\nu = c/2L$, where L is the resonator length. In our case this length was L = 130 cm and the separation was $\Delta\nu = 115$ MHz. The gain band of this (and other gas lasers) was known to be equal to the Doppler width of the laser transition, which was about 900 MHz [50, 51]. In the TEM_{n00} case the number of longitudinal modes in the radiation was equal to the number of times that $\Delta\nu$ could be fitted into the Doppler width: in our case about seven modes were emitted. The multimode nature of the radiation was not a serious disadvantage in square-law detection. The resultant signal represented the combined effect of the signals produced by each mode.

Measures were taken to ensure stabilization of the working conditions. The laser was isolated from vibrations in the following manner. A Duralumin plate was placed on two massive beams which projected from the main wall of the laboratory. Vacuum rubber spacers, 20 mm thick, separated the plate from the beams. The plate carried a wooden box containing sand of uniform grain size of the type used in casting. Layers of sand, 5 cm thick, alternated with layers of felt, 2 cm thick. A second Duralumin plate was placed on the last layer of felt and the laser was positioned on this plate. The laser was not thermostatted and therefore special measures were taken to ensure that the temperature was constant in the laboratory.

Solutions were placed in a cylindrical cuvette. The liquids used to make up these solutions were purified by a team of chemists in the Optical Laboratory of the Lebedev Physics Institute.*

Light scattered at some selected angle was selected by a lens ($f = 70$ mm) and focused onto the photocathode of a photomultiplier. A diaphragm with an aperture of about 1.5 mm diameter was placed in front of the photocathode. The aperture diameter and the position of the diaphragm were selected by trial and error in such a way as to achieve the strongest beat signal between the axial laser modes at a frequency of 115 MHz (this signal was examined with a spectrum analyzer of the S4-5 type).

It was not possible to decide a priori which of the photomultiplier types would be more suitable in optical heterodyning. It seemed that the photomultiplier which would be more sensitive to the constant component of the current would also have a higher sensitivity in respect

---

*The author is grateful to L. A. Novikova and L. V. Morozova for purification of some of the liquids used.

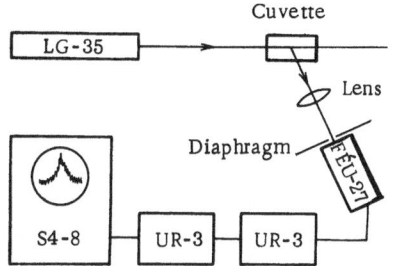

Fig. 2. Apparatus used in square-law detection of the scattered light.

of the alternating component. However, practice showed that this was not correct. A photomultiplier was selected from the following Soviet instruments: FÉU-27, FÉU-38, and FÉU-79. The selection criterion was the strongest signal of the 115 MHz beats between the axial modes. About 15 photomultipliers were tried. It was found that the strongest beat signal was produced by one of the FÉU-27 multipliers. The sensitivity of the cathode of this photomultiplier, measured with a source of light whose color temperature was 2850°K, was 48 μA/lm and the sensitivity threshold was 2.6×10^{-11}/lm. The sensitivity could be expressed in terms of watts at $\lambda = 6328$Å bearing in mind that a monochromatic beam of $\lambda = 5560$Å of 1 W power corresponded to a light flux of 683 lm [52]. When the last figure was multiplied by the ratio of the luminous efficiencies at $\lambda = 6328$ and 5560Å [52], it was found that 1 W of the $\lambda = 6328$Å radiation corresponded to 227 lm. Hence, we found that the sensitivity of the cathode of the selected photomultiplier was $\sim 10^{-13}$ W.

It was also convenient to know the sensitivity of the photocathode in electron/photon units in order to check that the 1% conversion coefficient assumed in the estimates was indeed achieved. A current of 48 μA was produced by $\sim 3 \times 10^{14}$ electrons. A flux of 1 lm consisted of $\sim 1.5 \times 10^{16}$ photons of $\lambda = 6328$Å wavelength. Thus, the coefficient of conversion of photons into electrons amounted to $\sim 2\%$.

The photomultiplier load was 430 Ω. The constant component of the current was measured by a microammeter in the load circuit.

The alternating signal produced by the photomultiplier was amplified by two series-connected wide-band amplifiers of the UR-3 type (pass band from 1 kHz to 200 MHz). The total gain achieved in these amplifiers was $\sim 10^2$ in respect of the voltage. The amplified signal was applied to an S4-8 spectrum analyzer which could measure signals in the frequency range from 3 kHz to 30 MHz.

The sensitivity threshold of the recording system was estimated as follows. The threshold voltage of the spectrum analyzer was 10^{-4} V. Hence, the voltage across the photomultiplier load had to be at least 1 μV. Such a voltage produced a current of 2.5×10^{-3} μA. The photomultiplier gain was $\sim 10^6$. Hence, the current produced by the photocathode was 2.5×10^{-9} μA. This gave a sensitivity threshold of $\sim 2 \times 10^{-13}$ W. Thus, the sensitivity threshold of the recording system was matched to the sensitivity of the photocathode of the multiplier.

The estimates given above were made for the constant component of the photomultiplier current. If it was assumed that the alternating component of this current was 10% of the constant component (which was not too optimistic), it was found that the sensitivity threshold of this spectrometer as a whole was $\sim 10^{-12}$ W.

The half-width of the transfer function of the spectrometer was measured under conditions of square-law detection of the laser radiation (Figs. 3 and 4). A calibrated signal of 12.6 ± 0.5 kHz was produced by a ZG-14 oscillator. The resolving power of the spectrometer was found to be $\sim 4 \times 10^{10}$. The linearity of the frequency response of the analyzer screen was

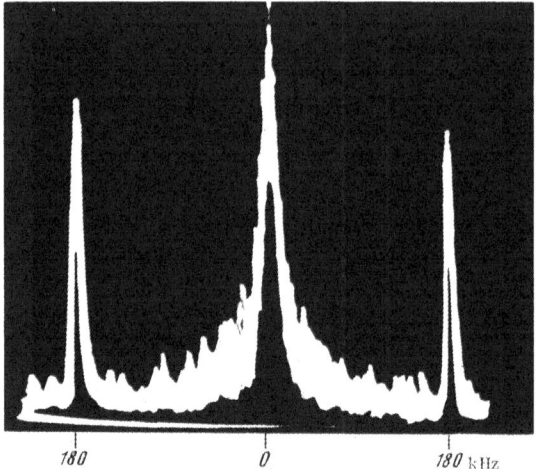

Fig. 3. Signal on the screen of a spectrum analyzer obtained in square-law detection of the
laser radiation. The spectrum includes a calibration signal of 180 kHz frequency.

Fig. 4. Signal on the screen of a spectrum analyzer obtained under the same conditions as
Fig. 3, except for the insertion of an opaque
screen in the laser beam. The spectrum includes calibration signals of 20 kHz (produced
by a ZG-14 oscillator) and 180 kHz frequencies.

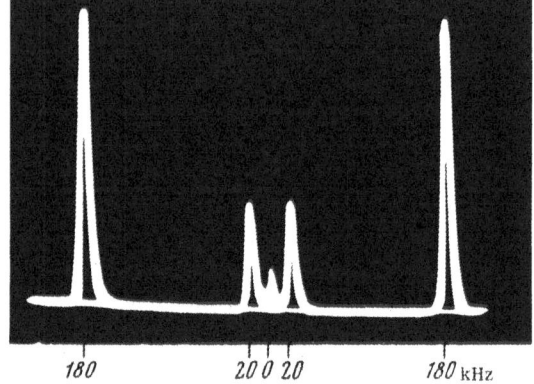

checked by means of calibrated signals from the ZG-14 oscillator and from a GSS-6 standard-
signal generator. The linearity of the circuit in respect of the voltage was verified by applying
calibrated signals from the ZG-14 oscillator to the inputs of the UR-3 amplifiers (the calibration was performed with a vacuum-tube voltmeter).

The noise spectrum displayed on the analyzer screen was photographed. An analysis of
this spectrum was performed with the aid of a magnifier. Selected frames were projected onto
a sheet of graph paper with a millimeter scale and the half-width was measured on this sheet.

The experiments were carried out in darkness. Numerous black-paper screens were
used to protect the photomultiplier from reflected or scattered laser radiation.

§2. Selection of Investigated Substances

If we combine Eqs. (I.16) and (I.17) with the van't Hoff law [44]

$$p = \frac{ckT}{v}, \tag{II.1}$$

where k is the Boltzmann constant and v is the molecular volume of the solvent, we obtain the
following expression for the coefficient of scattering at an angle of 90° ($R_{c,90}$) as a result of
fluctuations of the concentration in binary solutions:

$$R_{c,\,90} = \frac{2\pi^2}{\lambda^4}\, n^2 \left(\frac{\partial n}{\partial c}\right)^2_{\rho,\,T} \frac{cM}{N_A \rho}\,. \tag{II.2}$$

TABLE 1

CS₂, wt. %	n_{H_α}	CS₂, wt. %	n_{H_α}
0.000	1.35458	71.137	1.50931
13.245	1.37333	83.283	1.54325
29.326	1.40047	86.892	1.55692
40.329	1.42182	100.000	1.61440
51.902	1.44819		

Here, n is the refractive index of the solution; N_A is Avogadro's number; M is the molecular weight of the solvent; ρ is the density of the solvent. Equation (II.2) applies to ideal dilute solutions. However, we shall use it only for estimates and not for any exact measurements of quantity such as Avogadro's number.

Large values of R_c can be obtained by selecting the components of the solution so as to ensure the greatest possible difference between their refractive indices because the dependence n(c) of most of the solutions is nearly linear. One of the universal solvents with a high refractive index is carbon disulfide. Table 1 lists the values of the refractive index of acetone–carbon disulfide solutions (these values were recalculated on the basis of data given in [63] for the red line of hydrogen, H_α, of $\lambda = 6563$Å).

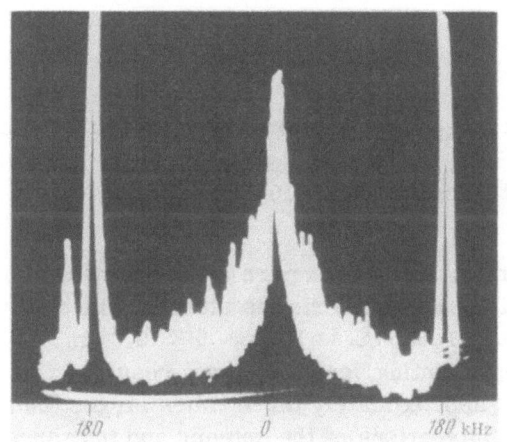

Fig. 5. Signal on the screen of a spectrum analyzer obtained in recording the light scattered at an angle of 90° in a solution of 10 wt.% acetone in carbon disulfide.

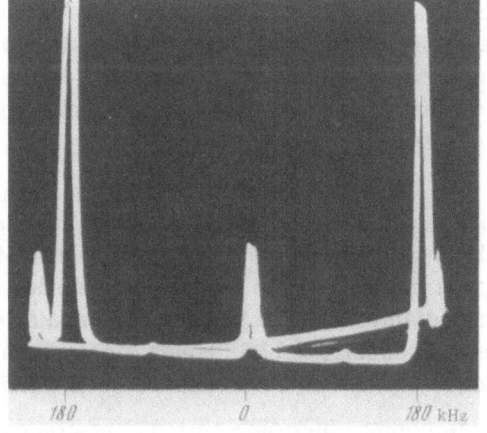

Fig. 6. Signal on the screen of a spectrum analyzer obtained when the photomultiplier cathode is not illuminated. This signal represents the output of the first heterodyne stage of the spectrum analyzer at zero frequency.

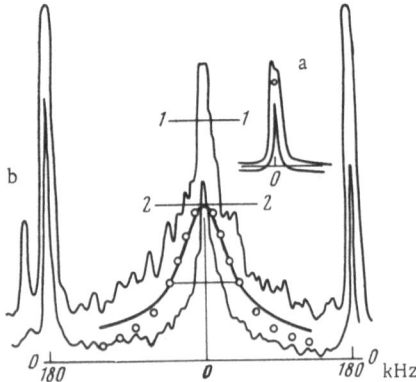

Fig. 7. Analysis of the signals shown in Figs. 5 and 6 (a) and of a stray signal (b). The separation between lines 1-1 and 2-2 is equal to the peak intensity of the signal a; the continuous curve represents the scattered-light signal averaged with respect to noise; the points represent a dispersion curve calculated using the measured half-width and the peak intensity; 0-0 is the zero line.

Fig. 8. Signal on the screen of a spectrum analyzer obtained for the light scattered at an angle of 111° in a solution of 10 wt.% acetone in carbon disulfide (a represents the signal obtained when the photocathode is not illuminated).

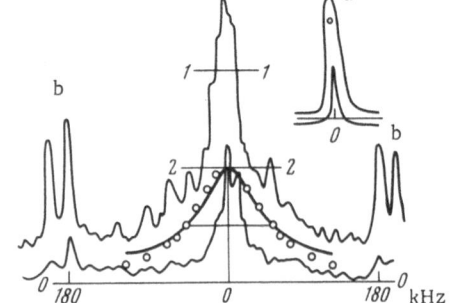

We shall now estimate the scattering coefficient $R_{c,90}$ for a solution containing 10 wt.% of C_3H_6O. We shall need the following parameters: $n \sim 1.6$; $\partial n / \partial c \sim 0.5$; $M_{CS_2} = 76$; $M_{C_3H_6O} = 58$; $c \sim m_{C_3H_6O} M_{CS_2} / m_{CS_2} M_{C_3H_6O} = 0.13$; $\lambda = 6563 \text{Å} = 66 \times 10^{-6}$ cm; $N_A = 6 \times 10^{23}$ g/mole; $\rho_{CS_2} = 1.26$ g/cm³.

An estimate of this kind gives $R_{c,90} \sim 9 \times 10^{-6}$ cm⁻¹. Let us compare this value with the coefficient of molecular scattering in benzene, for which such scattering is strong. The coefficient of molecular scattering in benzene [1], reduced to $\lambda = 6563\text{Å}$, is $R_{\delta,90} \sim 6 \times 10^{-6}$ cm⁻¹. We note that for benzene $I_S \sim I_B$. Thus, the coefficient representing scattering as a result of fluctuations of the concentration in the selected solution is approximately three times larger than the coefficients representing scattering as a result of fluctuations of the entropy and pressure in benzene. Consequently, in spite of the fact that the whole thermal scattering spectrum (Fig. 1) reaches the photocathode in square-law detection of the light scattered by a solution, the signal arriving at the spectrum analyzer is dominated by the contribution of the concentration scattering because the spectral density of this scattering is about 10^2 times greater than the spectral density of the entropy scattering and approximately 10^4 times greater than the spectral density of the Brillouin scattering.

We can now estimate the power of the scattered light which is analyzed in our spectrometer. A scattering volume of about 4 mm diameter is reduced slightly (the magnification is $\beta \sim 1$) in the process of imaging by a lens of 35 mm diameter and a focal length of 70 mm. This image is projected onto the photomultiplier cathode through the diaphragm with an aperture of 1.5 mm diameter. The power of the radiation reaching the photocathode is $\sim 6 \times 10^{-11}$ W. This value is approximately 60 times higher than the sensitivity threshold of the spectrometer and, consequently, it can be measured quite readily.

In direct measurements the constant component of the photomultiplier current was 1-3

μA when the applied voltage was 1.7-1.8 kV. Consequently, the total radiation power was $\sim 10^{-10}$ W, in good agreement with the estimate given above.

We investigated also solutions of bromoform ($CHBr_3$) in normal propyl alcohol ($n-C_3H_8O$). These solutions also exhibited strong scattering of light as result of fluctuations in the concentration. Moreover, the values of the diffusion coefficient had been determined earlier by physicochemical methods for a solution with a $1M$ concentration of bromoform. This made it possible to compare the values of the diffusion coefficients found by different methods.

§3. Results of Measurements and Discussion

A typical spectrum of the photocurrent noise, obtained in the spectral analysis of the light scattered by the investigated solutions, is shown in Fig. 5. The signal observed under the same conditions but in the absence of the laser beam is given in Fig. 6. In analysis of the spectrograms the intensity at the maximum of the signal shown in Fig. 6 was subtracted from the intensity at the maximum of the signal in Fig. 5 and the half-width was measured at midamplitude of the resultant curve at zero frequency. Spectrograms obtained as a result of such analysis are shown in Figs. 7 and 8.

The low value of the signal-to-noise ratio made it difficult to carry out these measurements. It is evident from our spectra that this ratio was of the order of unity, in agreement with the estimates given in Chap. I. The signal-to-noise ratio could be improved by the use of radiation detectors which would be more sensitive to the alternating component of the current.

Figure 7 shows the noise envelope taken from Fig. 5. It also gives the measured half-width, the dispersion curve calculated from the measured half-width and the intensity at the maximum, and the signal obtained by visual averaging. The position of the zero-frequency line was found by averaging noise in that part of the spectrum which was separated by five or six half-widths of the signal from the zero (central) point. We can see that within the half-width the signal follows closely the dispersion curve but considerable departures from this curve are observed in the wings. This may be due to the presence of the calibration signal and of strays or due to some other distorting factors.

The half-width was found ignoring the departures from the dispersion curve: the half-width of the photocurrent spectrum was divided into two.

When the half-width was recalculated taking account of the transfer function, it was found to depend on the scattering angle. The angular dependence of the half-width of the spectrum of light scattered by a solution of 10 wt.% acetone in carbon disulfide is given below:

θ, deg	45	90	111	135
$\Delta\nu$, kHz . . .	19.4$\pm$4.2	32.4$\pm$2.7	39.7$\pm$4.1	63.2$\pm$13.1

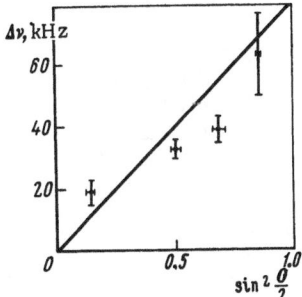

Fig. 9. Angular dependence of the half-width of the central component in the spectrum of the thermal scattering of light in a solution of 10 wt.% acetone in carbon disulfide (room temperature).

The scattering angle was measured to within 1-2°. The error quoted in the above table for the half-width was the random experimental error. The true error could be greater because of the presence of a systematic error associated with inaccurate determination of the intensity at the maximum, the position of the zero line, and errors committed in the averaging of noise.

Figure 9 shows the dependence of $\Delta\nu$ on $\sin^2(\theta/2)$. It was a priori assumed that this dependence was linear, in accordance with Eq. (1). The straight line plotted in Fig. 9 was obtained by the least-squares method. The arithmetic mean deviation of the experimental points from the straight line was ~30%. Thus, the angular dependence of the half-width of the scattering line obtained in our study should be regarded simply as a qualitative confirmation of the theory.

The measured values of the half-widths of the scattering spectra of solutions of bromoform in n-propanol and the interdiffusion coefficients calculated from these spectra are listed in Table 2. This table gives also the diffusion coefficient of an acetone—carbon disulfide solution, calculated from the slope of the straight line in Fig. 9.

The data in Table 2 and the angular dependence $\Delta\nu(\theta)$ described above were corrected for the systematic error of the earlier measurements [49]. This error was due to inaccurate averaging of noise at the zero frequency. It is evident from Table 2 that the diffusion coefficient deduced from the scattering spectrum is 1.9 times smaller than the diffusion coefficient deduced by a physicochemical method. This difference may be due to possibility of a 90% error in the present measurements and due to errors in the physicochemical measurements [54].

It is known that physicochemical measurements of the diffusion coefficients are very difficult [55, 56]. In these methods the change in the concentration during a given time interval is determined in a volume of the solution in contact with the solvent. Since the diffusion coefficients of solutions are small (they are usually of the order of 10^{-5} cm²/sec), the time interval in question must be long. During this time the solution must be carefully thermostatted because otherwise convection will result in mixing of the solvent and the solution. Moreover, it is necessary to isolate the apparatus from all possible vibrations. Obviously, it is difficult to determine the true diffusion coefficient of a solution under these conditions. If the constancy of the temperature is insufficient or if the isolation from vibrations is incomplete, the diffusion coefficients of solutions are overestimated.

In optical heterodyning the diffusion coefficient can be determined practically instantaneously. Although the experimental error in the measurements of the diffusion coefficient re-

TABLE 2. Half-Width of the Central Component in the Scattered-Light Spectra of Solutions and Interdiffusion Coefficients

Components of solutions		Concentration of B	n_{H_α}	$\delta\nu_c = \frac{\delta\omega_c}{2\pi}$, kHz $\theta = 90°$	$D \cdot 10^5$, cm²/sec
A	B				
n- C_3H_8O	$CHBr_3$	1 M	1.398	49.6±3.1	0.38 ±0.04 0.71*
		2 M	1.418	69.1±15.2	0.52 ±0.14
		3 M	1.453	79.7±9.9	0.57 ±0.13
		4 M	1.548	68.9±3.3	0.43 ±0.04
CS_2	C_3H_6O	10 wt. %	1.571	$\frac{\Delta\nu}{\sin^2(\theta/2)} = 80.2$ kHz	0.26 ±0.07

* Value of the diffusion coefficient taken from [54].

ported here is high, there is no doubt that improvement of the apparatus would make it possible to measure these coefficients much more accurately.

CHAPTER III

THERMAL AND STIMULATED BRILLOUIN SCATTERING, VELOCITY OF HYPERSOUND, AND DISPERSION OF THIS VELOCITY IN AQUEOUS SOLUTIONS OF TERTIARY BUTYL ALCOHOL

§1. Some Features of the Propagation of Ultrasound in Aqueous Solutions of Tertiary Butyl Alcohol

Aqueous solutions of tertiary butyl alcohol are interesting because of the special features of the propagation of ultrasound in these solutions. Burton [26] measured the velocity and absorption of ultrasound in solutions of tertiary butyl alcohol in water. These measurements were carried out in the frequency range $(5-25) \times 10^6$ Hz at 27°C. They showed that the velocity and the absorption had definite maxima at 4.5 and 11 mol.% of alcohol, respectively. The ratio α/f^2 was 38.5×10^{-15} sec^2/cm at the maximum of the absorption curve (dashed curve in Fig. 13); here, α is the total amplitude absorption coefficient of ultrasound. This value is two orders of magnitude greater than the value calculated for the same concentration from a hydrodynamic formula in which only the shear viscosity is allowed for.

Burton suggested that the additional absorption is due to the formation of liquid crystals. However, later x-ray diffraction studies did not confirm this hypothesis. Measurements of the refraction of the solutions, which were also carried out by Burton, demonstrated that the refraction was a linear function of the concentration. This behavior was typical of nonassociated solutions and, therefore, the additional absorption of ultrasound was not due to the intermolecular interaction.

Nomoto [57, 58] suggested a molecular model for the absorption of ultrasound in aqueous solutions of alcohols. He postulated the existence of associated complexes of water molecules and of isolated water and alcohol molecules in these solutions. When an acoustic wave passes through a solution, the higher pressures dissociate complexes into single molecules. Therefore, alcohol molecules diffuse in regions where the concentration of water molecules is high. Such diffusion is accompanied by an irreversible increase in the entropy and, consequently, it results in the absorption of ultrasound. The Nomoto theory includes parameters which are difficult to find from independent measurements. For example, according to Nomoto, water complexes should consist of about 800 molecules each. Naturally, it is difficult to check a theory which contains such parameters. Romanov and Solov'ev [59, 60] attributed absorption of sound in solutions to the relaxation of fluctuations of the concentration and they derived a phenomenological formula for the absorption of sound. The application of this formula to acetone—water solutions gave results in good agreement with the experimental data. The formula has not yet been checked against the results for aqueous solutions of tertiary butyl alcohol because many of the thermodynamic parameters, needed in this formula, have not yet been measured.

The fine structure of the Rayleigh line can be observed if the absorption of hypersound α in a wavelength Λ is small, i.e., if $\alpha\Lambda \ll 1$. In observations of the Brillouin scattering at an angle of 90° the wavelength of hypersound is $\Lambda \sim \lambda/2$ (λ is the wavelength of the exciting radiation) and the frequency of hypersound is $f \sim 10^{10}$ Hz. Hence, if $\alpha/f^2 = 38.5 \times 10^{-15}$ sec^2/cm, we find that $\alpha\Lambda \sim 240$. This means that the fine structure can be observed when the absorption decreases strongly, i.e., when it relaxes.

If the large value of α/f^2 of solutions is due to a high value of the bulk viscosity and the frequency dependence of the absorption of sound can be described by a simple relaxation theory with just one bulk-viscosity relaxation time τ, we find that [1, 27]

$$\frac{\alpha}{f^2} = \frac{\alpha_\eta}{f^2} + \frac{\alpha_{\eta'}}{f^2} = B + \frac{A}{1 + \left(\frac{f}{f_C}\right)^2},$$ (III.1)

where

$$\frac{\alpha_\eta}{f^2} = B = \frac{8\pi^2}{3\rho v_0^3}\,\eta, \quad A = \frac{\pi}{v_0^3 f_C}[v_\infty^2 - v_0^2], \quad f_C = \frac{1}{2\pi\tau}.$$ (III.2)

Here, α_η and $\alpha_{\eta'}$ are the amplitude absorption coefficients due to the shear (η) and the bulk (η') viscosities: v_∞ and v_0 are the values of the velocity of sound for $f \to \infty$ and $f \to 0$, respectively; ρ is the density; f_C is the critical relaxation frequency of the bulk viscosity.

If Eqs. (III.1) and (III.2) are applicable, a positive dispersion of the velocity of sound should be observed. The dispersion $\Delta v/v$ is given by [1]

$$\frac{\Delta v}{v} = \frac{v}{4\pi^2\tau}\frac{\alpha_\eta}{f^2}\left(\frac{\alpha}{\alpha_\eta} - 1\right) \approx \frac{v}{4\pi^2\tau}\frac{\alpha}{f^2},$$ (III.3)

where $v = 1/2(v_\infty + v_0)$ and it is assumed that $\alpha \gg \alpha_\eta$. Burton [26] found no dispersion in the velocity of sound between 5×10^6 and $f_{max} = 25 \times 10^6$ Hz. This means that $\tau < 1/2\pi f_{max} \approx 6 \times 10^{-9}$ sec and, consequently, if Eqs. (III.1)–(III.3) are applicable, we should expect a dispersion $\Delta v/v \approx 2.5\%$ in the region of $v \approx 1.5 \times 10^5$ cm/sec. If the relaxation time in the solutions in question were equal to the relaxation time in pure liquids, i.e., $\tau \sim 10^{-10}$ sec, the dispersion of the velocity of sound should be enormous $(\sim 150\%)$.

The true dispersion of the velocity of sound can only be determined experimentally [24].

§2. Apparatus and Measurement Method

The velocity of hypersound v_h was determined at $\sim 5 \times 10^9$ Hz from the shift of the Brillouin components in the spectra of the thermal and stimulated scattering. This was done using Eq. (I.24) rewritten in the form

$$v_h = \frac{\Delta v c}{2n v_L \sin\frac{\theta}{2}} = \Delta v \Lambda.$$ (III.4)

The velocity of ultrasound was measured at 2×10^6 Hz. All the measurements were carried out at 21°C. The thermal Brillouin scattering spectra were determined using an interferometer described earlier [6, 61]. The scattering was excited by a neon–helium laser ($\lambda = 6328$Å) of 20-25 mW power.

Cuvettes used in this study had plane-parallel windows in the optical paths. Carefully purified tertiary butyl alcohol and twice-distilled water were mixed in the required proportions and poured into a cuvette through a glass filter No. 4. The alcohol melted at 25.5°C. However, our study showed that well-purified alcohol was slightly supercooled and it did not solidify for a long time when it was poured into a dust-free cuvette. Thus, we were able to record the scattering spectra of pure tertiary butyl alcohol at 21°C.

Fig. 10. Apparatus used in investigations of the stimulated
scattering of light. M is a plane mirror with a dielectric
coating and a reflection coefficient of $\sim 100\%$; C_1 and C_2 are
cuvettes containing cryptocyanine solutions; R_1 and R_2 are
ruby crystals; S is a mode selector; A is an attenuator; l
is a lens; V is a cell containing the solution under investigation; F is a red filter; F-P is a Fabry–Perot interferometer; L is an objective ($f = 1200$ mm); P is a photographic film.

The light scattered at an angle of $\theta = 90 \pm 0.2°$ was analyzed with a Fabry–Perot interferometer. The interferometer mirrors had multilayer dielectric coatings with a reflection
coefficient R $\sim 98\%$. The ring separating the mirrors was t = 8 mm thick. The dispersion
amounted to 0.625 cm^{-1}. The focal length of the camera objective was 600 mm. Spectra were
photographed on 10N-1000 panchromatic film. The exposure lasted 2-3 h.

The linear separations between the fine-structure components of the Rayleigh line were
measured with an IZA-2 comparator. The frequency shift of the Brillouin components was calculated by the off-center method described in [1, 62]. The values of the refractive index needed
in the calculation of the velocity of hypersound were determined with an IRF-22 refractometer.
The velocities were found by averaging eight independent measurements. The error in the determination of the velocity of hypersound was 0.5-0.7%.

The stimulated Brillouin spectra were obtained using apparatus shown schematically in
Fig. 10. The excitation was provided by giant pulses produced by a ruby laser, as was done in
earlier investigations [63-65].

The laser radiation pulses were time-scanned with an FÉK-09 coaxial photocell (time
resolution $\sim 5 \times 10^{-10}$ sec) and an I2-7 oscillograph (time resolution $\sim 10^{-9}$ sec). Such scanning
demonstrated that, depending on the concentration of cryptocyanine in alcohol solutions used
for Q switching, the half-width of the laser pulses ranged from 10 to 20 nsec. The laser radiation energy was measured with a calorimeter and it amounted to ~ 1 J. Consequently, the power
of ~ 10 nsec pulses was ~ 100 MW. When these laser pulses were focused by a lens of $f \approx 4$ cm,
the power in the scattering region was considerably higher than the threshold necessary for the
stimulated Brillouin scattering in most liquids. Local multiple scattering was also observed,
i.e., the Brillouin component of the stimulated scattering generated a new stimulated Brillouin
component, etc. The number of the stimulated Brillouin lines could be so large as to make it
difficult to interpret the spectra. Moreover, it was clearly evident that in order to determine
the parameters of a given substance from its stimulated scattering spectrum it was necessary
to use laser radiation powers close to the threshold because the mutual influence of the various
nonlinear effects could not be allowed for easily at high power levels [66].

In view of this the laser power radiation was attenuated by a pile of glass plates A down
to ~ 20 MW and the radiation was focused into the solution-filled cuvette V by a long-focus lens
l ($f = 140$ mm). Feedback was observed between the scattering medium and the laser. Consequently, the Brillouin radiation scattered through 180° was recorded before it was amplified in
the laser.

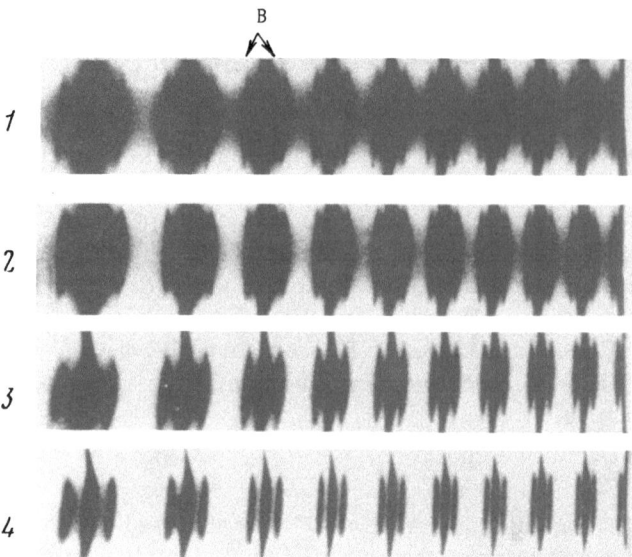

Fig. 11. Spectra of the thermal Brillouin scattering (B) in aqueous solutions of tertiary butyl alcohol at t = 21°C. Alcohol concentration (molar fractions): 1) 0.045; 2) 0.110; 3) 0.300; 4) 1.000.
Spectral range of the interferometer 0.625 cm^{-1}.

The stimulated Brillouin scattering spectrum was determined with a Fabry—Perot interferometer F-P (spectral range 0.625 cm^{-1}) and was focused on a photographic film P by an objective L with a focal length of 1200 mm.

The laser radiation spectrum was recorded before the main measurements. This was

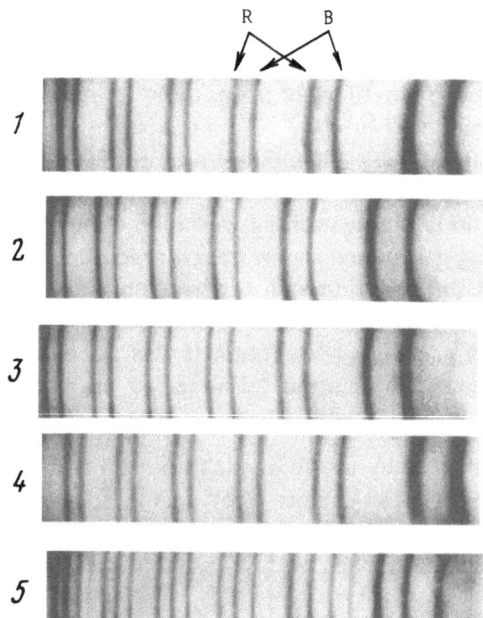

Fig. 12. Spectra of the stimulated Brillouin scattering (B) in aqueous solution of tertiary butyl
alcohol at t = 21°C. Alcohol concentration (molar
fractions): 1) 0.000; 2) 0.45; 3) 0.110; 4) 0.300;
5) 1.000. R denotes the ruby lines.

done in order to study the mode structure of the laser beam. The width of the laser line was found to be ~ 0.05 cm^{-1}. The stimulated Brillouin scattering spectra were interpreted in the same way as the thermal scattering spectra discussed above. In those cases when local multiple scattering lines were observed, the measurements were restricted to the positions of the laser line and the first Stokes component in the interference pattern.

The velocity of ultrasound was measured using apparatus described in [67, 68].

All the hypersonic and ultrasonic measurements were carried out on the same solutions.

§3. Results and Discussion

Typical thermal and stimulated Brillouin scattering spectra of aqueous solution of tertiary butyl alcohol are shown in Figs. 11 and 12 (plate). The results of the measurements are given in Tables 3 and 4 and in Figs. 13 and 14.

The dispersion of the velocity of sound was deduced from the thermal Brillouin scattering spectra. The maximum dispersion corresponded to the absorption peak and it amounted to $7.0 \pm 1.1\%$. This was a large value for low-viscosity liquids, i.e., liquids with a static shear viscosity $\eta \sim 1$ cP, such as the solutions being investigated. The maximum dispersion of the velocity of sound is known to be exhibited by carbon tetrachloride and its value is 11% [1]. According to the theory given in [27], these values of the dispersion indicate that the solutions being investigated have bulk-viscosity relaxation times $\sim 10^{-9}$ sec. The relaxation times τ calculated from the observed dispersion by means of Eq. (III.3) are listed in Table 4. This table gives also the critical relaxation frequencies $f_C = 1/2\pi\tau$. The results obtained show that an

TABLE 3. Refractive Indices ($n_{6328Å}$ and $n_{6943Å}$) of Aqueous Solutions of Tertiary Butyl Alcohol at t = 21°C

Alcohol conc., molar fractions	n_{6328A}	n_{6943A}	Alcohol conc., molar fractions	n_{6328A}	n_{6943A}
0.000	1.3315	1.3294	0,300	1.3735	1.3710
0.045	1.3414	1.3393	1.000	1.3865	1.3845
0.110	1.3584	1.3561			

TABLE 4. Velocities of Ultrasound v_u and Hypersound v_h and Some Parameters of Aqueous Solutions of Tertiary Butyl Alcohol (t = 21°C) Deduced from These Velocities

Alcohol conc., molar fractions	$f_h^s \cdot 10^{-9}$, Hz	$f_h^t \cdot 10^{-9}$, Hz	v_h^s, m/sec	v_h^t, m/sec	v_u, m/sec	$\frac{\Delta v}{v}$, %	$\frac{\alpha}{f^2} \cdot 10^{15}$, sec^2/cm [26]	$\tau \cdot 10^{9}$, sec	$f_C = \frac{1}{2\pi\tau} \cdot 10^{-9}$, Hz
0.000	5.61	4.42	1464±16	1488.5±1.3a	1488.5±1,5	—	0.25	—	—
0.045	6.33	5.06	1642±7	1686±20	1622±3	3.9±1.4	10.7	11	0.145
0.110	6.09	4.94	1559±10	1627±15	1518±2	7.0±1.1	38.5	22	0.072
0.300	5.33	4.24	1351±10	1383±13	1337±5	3.3±1.3	17.0	17	0.094
1.000	4.64	3.66	1164±6	1181±8	1150±2	2.7±0.9	5.8	6	0.265

Subscripts: s is the stimulated scattering, t is the thermal scattering, a are the results of Tiganov [69].

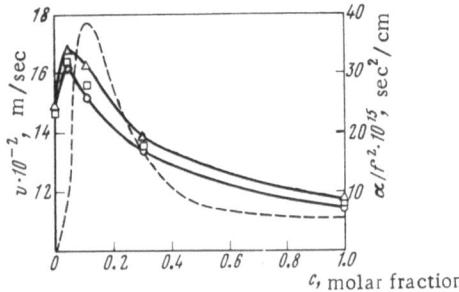

Fig. 13. Velocities of ultrasound (circles) and of hypersound deduced from the spectra of the stimulated (squares) and thermal (triangles) Brillouin scattering in aqueous solutions of tertiary butyl alcohol at t = 21°C. The dashed curve represents the absorption of ultrasound measured by Burton [26] in the same solutions at t = 27°C.

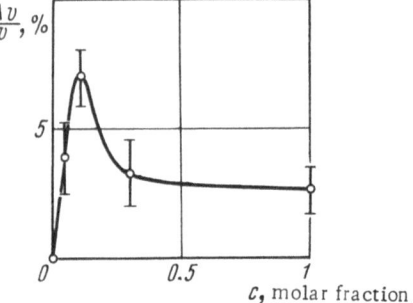

Fig. 14. Dispersion of the velocity of sound Δv/v in aqueous solutions of tertiary butyl alcohol.

increase in the absorption of ultrasound in aqueous solutions of tertiary butyl alcohol is accompanied by an increase in the dispersion of the velocity of sound and of the relaxation time (the critical relaxation frequency decreases).

These results show that the solutions under investigation are not Kneser liquids [70, 71]. The addition of an impurity to liquids of this type reduces the absorption of sound and the relaxation time [71]. The fall of the relaxation time is attributed to the easier exchange of the energy between the internal and external degrees of freedom of molecules in the presence of an impurity [70]. Romanov and Solov'ev [59] derived a phenomenological formula which describes the absorption of sound in solutions. According to this formula,

$$\frac{\alpha}{f^2} \propto \frac{1}{D}, \tag{III.5}$$

where D is the diffusion coefficient.

It follows from Eq. (III.5) that the absorption of sound may increase because of a reduction in the diffusion coefficient D. Such a behavior of the quantities α/f^2 and D is indeed observed near the critical solution point. As is known, this critical point is two-dimensional: a solution separates into its components at some definite temperature and concentration. Therefore, the maximum of the absorption of ultrasound observed at 21°C at the concentration of 10 mol.% (0.1 molar fractions) indicates that under these conditions the solution is close to its critical point.

The velocity of ultrasound in aqueous solution of tertiary butyl alcohol was also measured* also at 2.55 × 10⁶ Hz at 27.5°C. These measurements were combined with those carried out at 21°C at 2 × 10⁶ Hz in a calculation of the temperature coefficient of the velocity of ultra-

---

*These measurements were carried out by S. V. Krivokhizha, who kindly supplied us with the results.

sound $\Delta v_u / \Delta t$ in aqueous solutions of tertiary butyl alcohol. The results obtained were as follows:

Alcohol conc., molar fraction	0	0.045	0.11	0.3	1.0
$\Delta v_u / \Delta t$, m·sec^{-1}·deg^{-1}	+1.2(±1)	+3.8(±0.7)	+0.15(±0.5)	−0.7(±1)	−5(±0.5)

It is evident from the data given above that solutions of concentrations ranging from 0.11 to 0.3 molar fractions have temperature coefficients of the velocity of ultrasound which are close to or even equal to zero. Such media can have practical applications as acoustic modulators of light in which the acoustic wavelength must be stable and, consequently, it is necessary to ensure that the velocity of sound does not vary with the temperature of the medium in which the acoustic wave is traveling.

Measurements of the temperature coefficient of the velocity of ultrasound were of intrinsic interest but the aim was to interpret the data on the velocity of hypersound deduced from the stimulated Brillouin scattering spectra. We shall now consider these spectra in greater detail.

The stimulated Brillouin scattering in solutions was first observed, independently and at approximately the same time, by Barocchi, Mancini, and Vallauri [72], by Bespalov and Kubarev [73], and by Aref'ev, Starunov, and Fabelinskii [24]. There is no basic difference between the stimulated Brillouin scattering in pure liquids and in solutions and therefore we shall not discuss this phenomenon in greater detail (this was done earlier).

The stimulated Brillouin scattering spectra can be recorded very rapidly and the positions of the Brillouin components can be found very accurately. However, the velocities of hypersound deduced from these spectra are found to be systematically lower than the values obtained from the thermal scattering spectra.

It is evident from Table 4 that this difference amounts to 1.5–4%. Mash et al. [74] were the first to draw attention to this discrepancy: they attributed it to the heating of the scattering volume (they investigated samples with a negative temperature coefficient of the velocity of hypersound). Since the temperature coefficients of the velocity of sound in water and in solutions up to 0.11 molar fractions are positive (as discussed above), the mechanism suggested by Mash et al. cannot be of great importance in our case and the effect must be explained by a different mechanism.

The difference between the values of the velocity of hypersound deduced from the thermal (v_h^t) and the stimulated (v_h^s) spectra is due to the following effect. In our experiments the feedback between the scattered radiation and the laser was not suppressed and the scattering was studied in the forward direction. Therefore, the Brillouin components included a contribution from the back-scattered light ($\theta = 180°$) which was amplified in the laser, returned to the cuvette, and reached the interferometer. This amplification of the stimulated Brillouin scattered light in the laser could occur only at one of the frequencies of the natural modes of the laser. We know that $\Delta \nu \times 2L/c$ modes (L is the length of the optical resonator) can be fitted into the half-width of the emission line of ruby $\Delta \nu$. The stimulated Brillouin scattering line may lie in the interval between two modes. Then, depending on whether the mode closest to the Brillouin line has a higher or a lower frequency, the frequency of the stimulated Brillouin scattering will either increase or decrease.

Thus, the frequency of the stimulated Brillouin scattering line is indeterminate to the extent represented by the separation between the longitudinal modes of the laser resonator. This was first established experimentally by Brewer [47]. The mechanism suggested above explains why the velocity v_h^s cannot be equal to the velocity v_h^t but it does not explain why $v_h^s < v_h^t$.

Chaban [75] suggested why $v_h^s < v_h^t$. The physical basis of Chaban's explanation can be stated as follows. The number of phonons which are being created is greater in the region where the gain is higher. Since beams are focused by lenses, the gain is higher in the region where the optical path is longer, i.e., on the periphery of a focused beam. Thus, the forward-scattered light is effectively modulated by phonons which are traveling at an angle to the laser axis. Chaban found that the measured velocity of hypersound was equal to $v \cos \theta_0$, where θ_0 is the half-aperture of the focused laser beam. Hence, the difference between the velocities $\delta v/v$ should be $100(1 - \cos \theta_0) = 100 \cdot 2 \cdot \sin^2 \theta_0/2 \approx 50 \, \theta_0^2 \%$. In our experiments the laser beam was of ~10 mm diameter. For a lens of $f = 140$ mm we found that $\theta_0 \approx 5/140 \approx 0.035$ and $\delta v/v \sim 0.1\%$. This was considerably smaller than the discrepancy found experimentally. Thus, Chaban's hypothesis does not explain our results.

Morozov [66, 76] studied the stimulated Brillouin scattering in compressed nitrogen as a function of the intensity of the exciting radiation. This intensity was varied by the use of focusing lenses with different focal lengths. Morozov showed that the shift of the Brillouin components decreased with increasing intensity of the exciting radiation. However, he did not provide a definite explanation why this occurred.

Goldblatt and Litovitz [77] suppressed feedback between the laser and the cuvette in which scattering took place. This was done by the use of a quarter-wave plate. It was found that when single-mode laser radiation was employed (this radiation consisted only of longitudinal modes selected by a diaphragm of ~1 mm diameter placed within the resonator), the velocities of hypersound in water and in carbon disulfide were in agreement with the values calculated from the ultrasonic data.

Thus, one can specify experimental conditions under which reproducible values of the velocity of hypersound can be deduced from the stimulated Brillouin scattering spectra but it is still not clear why v_h^s is systematically smaller than v_h^t.

An interesting feature of the stimulated Brillouin scattering spectra obtained in the present study was the fact that the scattering threshold of tertiary butyl alcohol was higher than the thresholds of the solutions of this alcohol and of water (Fig. 12).

The stimulated Brillouin scattering in solutions can also be used, at a fixed scattering angle, to produce a tunable light source whose frequency can be varied within a narrow range. If the components of the solution are selected to differ as much as possible in the velocities of sound and one component is added gradually to the other, the frequency can be varied within 0.1 cm^{-1}. Attention was drawn to this point by Bespalov and Kubarev [73].

CHAPTER IV

FINE STRUCTURE OF THE RAYLEIGH LINE AND TEMPERATURE DEPENDENCE OF THE VELOCITY OF HYPERSOUND AND OF DISPERSION OF THIS VELOCITY NEAR THE CRITICAL SOLUTION POINT OF TRIETHYLAMINE—WATER SYSTEMS

§1. Some Features of the Scattering of Light near the Critical Point of Solutions

We have mentioned earlier that the intensity of light scattered by concentration fluctuations, I_c, increases strongly on approach to the critical point at which a solution separates into its components ("critical solution point"). The intensities of the Brillouin components of the Rayleigh triplet should not change greatly compared with I_c. According to the experimental results reported in [78] the thermal conductivity shows no anomalies near the critical solution point and the specific heat C_p increases slightly. This means that the width and intensity of the scattering line resulting from entropy fluctuations are not greatly affected.

The intensities of the Brillouin components of the Rayleigh triplet are given by the expression [1]

$$2I_B = I_0 \frac{\pi^2 V}{2\lambda^4 L^2} \left(\rho \frac{\partial \varepsilon}{\partial \rho} \right)_S^2 \beta_S kT (1 + \cos^2 \theta). \qquad (IV.1)$$

The adiabatic compressibility β_S is

$$\beta_S = \frac{1}{\rho v^2}. \qquad (IV.2)$$

The velocity of hypersound does not vary greatly near the critical solution point [79, 80]. Therefore, the intensities of the Brillouin components should not be greatly affected. It follows that the relative intensity of the Brillouin components in the Rayleigh triplet, $I_B / (I_c + I_S)$, should decrease strongly on approach to the critical point. For this reason the fine structure of the Rayleigh line of a solution near its critical point cannot be investigated with a mercury lamp. This applies also to studies in the vicinity of the critical points of pure substances. Therefore until recently there have been no published experimental data on the behavior of the Brillouin components near the critical point of solutions and in the critical regions of pure substances.

The development of neon–helium lasers has made it possible to determine for the first time the fine structure of the Rayleigh line near the critical solution temperature and to measure the temperature dependence of the velocity of hypersound [79 80]. The first investigation of the fine structure near the critical point of a pure substance (CO_2) was carried out by Gammon, Swinney, and Cummins [81], who used a gas laser. Ford, Langley, and Puglielli [82] also determined the fine structure of the Rayleigh line near the critical point of CO_2 and determined the temperature dependence of the width of the Brillouin components. Guberman and Morozov [83] studied the stimulated Brillouin scattering in CO_2 and determined the frequency shift of the Brillouin components in the vicinity of the critical point. The results reported in [81-83] are basically in agreement. Guberman and Morozov [83] found that the optical and acoustic losses near the critical point became so large that, under the conditions used in their experiments, the Brillouin scattering disappeared when the temperature approached the critical point to within 0.5 deg. Guberman and Morozov [83] were the first to use the stimulated Brillouin scattering in a study of the velocity of hypersound near the critical point. However, it is not clear to what extent the properties of a substance subjected to a strong optical field can depart from the critical values.

§2. Velocity and Absorption of Ultrasound near the Critical Solution Point of Triethylamine–Water Systems

Our measurements were carried out on a triethylamine–water solution containing 44.6 wt.% of triethylamine. These solutions had the lower critical temperature $t_C = 17.9°C$, which was convenient for measurements, and which had been studied thoroughly in the ultrasonic range [84, 85].

Chynoweth and Schneider [84] demonstrated that the velocity of ultrasound v_u of $f = 6 \times 10^5$ Hz frequency decreased on approach to the critical solution point of the triethylamine–water system (Fig. 17), whereas the absorption increased so that the ratio α/f^2 reached $\sim 10^{-12}$ sec²/ cm. According to Semenchenko and Zorina [86], the shear viscosity η of the triethylamine– water solutions increased by 15-20% and became ~ 4 cP at the critical point. It follows from Eq. (III.2) that the absorption of ultrasound due to the shear viscosity is given by $\alpha_\eta/f^2 \approx 0.4 \times 10^{-15}$ sec²/cm and, therefore, the observed absorption of ultrasound was practically all due to the bulk viscosity. A comparison of results of Chynoweth and Schneider [84] and those given by Sette [85], who measured the absorption of ultrasound in the same solution at t = 15°C in the frequency range 7.6×10^5-52.3×10^6 Hz, indicated that – in the approximation of a single relaxation time τ – the critical relaxation frequency of the bulk viscosity at t = 15°C was $f_C \approx 3 \times 10^6$ Hz. If the remaining absorption was represented by $\alpha/f^2 \approx 0.4 \times 10^{-15}$ sec²/cm, it was found that $\alpha\Lambda \approx 0.3 < 1$ (for the usual case of the scattering by hypersonic waves at an angle of 90°, for which $f \approx 5 \times 10^9$ Hz and $\Lambda \approx 3 \times 10^{-5}$ cm) and the fine structure of the Rayleigh line should be observed. In view of this, an investigation was made of the triethylamine–water solutions by the method described below.

§3. Experimental Method

The apparatus used in this investigation is shown schematically in Fig. 15. The excitation source was an OKG-12 neon–helium laser emitting at $\lambda = 6328$Å (output power 14 mW). The solution being investigated was poured at about 10°C into a glass cuvette of about 150 cm³ volume through a glass filter No. 4. The purity of the solution was checked by comparing its critical point with the critical temperature given in [84]. The critical solution point was determined visually as the temperature at which the solution became turbid. The cuvette had double windows and the space between the windows was evacuated. It also had a jacket through which water was circulated from a Höppler ultrathermostat. This thermostat was fed with tap water whose temperature was 7°C (the experiments were carried out in winter) and the water was heated in the thermostat. The temperature of the solution was kept constant to within at least ± 0.05 deg. The temperature was measured with a mercury thermometer with 0.1 deg scale divisions (it could be read to half a division with a magnifying glass). Thermal equilibrium was ensured by maintaining a given temperature for at least 4.5-5 h.

When a solution separated into its components, the homogeneity could not be restored simply by lowering the temperature. It was also necessary to stir the solution. This could be done by taking the cuvette out of the apparatus and shaking it. Such an operation had to be avoided because the spectra should be recorded with the cuvette exactly in the same position. Therefore, the temperature was maintained at 10-14°C continuously. The scattered light was

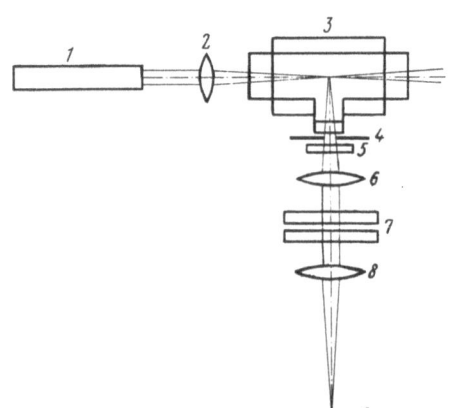

Fig. 15. Apparatus used in studies of the fine structure of the Rayleigh line in solutions near the critical solution point: 1) neon–helium laser; 2) lens ($f_1 = 100$ mm); 3) cell containing the solution under investigation; 4) diaphragm; 5) red filter; 6) objective ($f_2 = 210$ mm); 7) Fabry–Perot interferometer; 8) objective ($f_3 = 600$ mm); 9) photographic film.

observed at an angle of 90° with respect to the incident beam of the exciting radiation ($\theta = 90°$). The scattering angle was measured with a pentaprism to within $\pm 0.5°$.

The scattered light was decomposed into a spectrum with a Fabry–Perot interferometer (spectral range 0.5 cm^{-1}). It was necessary to ensure a high contrast of the interference pattern C, given by the formula [1]

$$C = \frac{(1+R)^2}{(1-R)^2},\qquad\qquad\text{(IV.3)}$$

which increased with increasing reflection coefficient R of the interferometer plates. On the other hand, the transmission of the interferometer had to be sufficiently high in order to reduce the exposure time to a reasonable value. Experience accumulated in our laboratory suggested that the best conditions were achieved when the reflection coefficient was R = 95%. Therefore, multilayer dielectric mirrors with this value of the reflection coefficient were used in the Fabry–Perot interferometer.

The preliminary measurements were carried out using a camera with a focal length $f_3 = 270$ mm. Subsequently, extensive control measurements were carried out using a camera with $f_3 = 600$ mm.

The interference pattern was photographed on an A700 photographic film. The exposure time near the critical point was 10-15 min when the camera with $f_3 = 600$ mm was used. The spectra were analyzed as described in Chap. III, §2.

The refractive index of the solution was determined at various temperatures with an IRF-22 refractometer.

Far from the critical point the velocity of hypersound was found by averaging the results of 8-10 independent measurements but near the critical point the results of 10-20 measurements were averaged to give a single value.

§4. Results and Discussion

Typical spectra are shown in Fig. 16. We can see that although the scattering at the frequency of the laser line is very strong near the critical solution point, the Brillouin components are still observed clearly. The relative error in the determination of the shift of the Brillouin components $\delta(\Delta\nu)/\Delta\nu$ is given below as a function of the temperature difference ($t - t_C$):

$(t - t_C)$, °C ...	8.9	6.9	4.9	2.9	1.9	1.4	0.9	0.6	0.3	0.2	0.1
$\dfrac{\delta(\Delta\nu)}{\Delta\nu}$, % ...	1.2	0.7	0.5	1.9	2.5	2.9	3.0	3.0	4.6	1.8	6.0

We can see that the precision of measurement of the shifts of the Brillouin components deteriorates beginning from about 3 deg C from the critical point and that in this range the er-

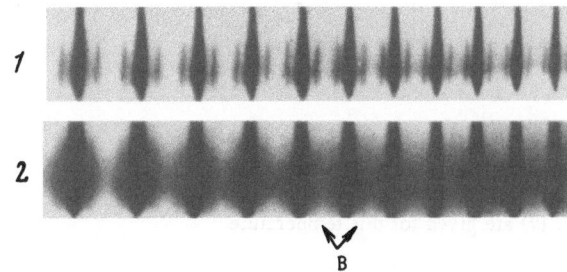

Fig. 16. Fine structure of the Rayleigh line of a 44.6 wt.% solution of triethylamine in water at 8.5°C (1) and 17.8°C (2). Spectral range of the interferometer 0.5 cm^{-1}.

ror amounts to a few percent. This is the precision that can be achieved in measurements of the positions of the Brillouin components of pure liquids when a mercury lamp is used as an excitation source [1].

The results of measurements of the refractive index n at λ = 6328Å, of the velocity of hypersound v_h at a frequency $f \sim 0.5 \times 10^{10}$ Hz, and of the dispersion of the velocity of sound $\Delta v/v$ in the investigated solution of triethylamine in water are presented in Table 5 and Figs. 17 and 18.

The results obtained show that the velocity of hypersound increases on approach to the critical solution point, which is very unusual for the behavior of this velocity near critical points.

We shall now interpret the results using the Mandel'shtam–Leontovich relaxation theory, which describes the propagation of sound in liquids. Surprisingly, the simple variant of this theory, which postulates a single relaxation time τ, yields results which are close to those obtained experimentally. For example, it is reported in [84, 85] that at t = 15°C the ratio α/f^2 is 1360×10^{-15} cm^2/sec for $f = 0.6 \times 10^6$ Hz and $\tau \approx 6 \times 10^{-8}$ sec. If we use the values and apply Eq. (III.3), we find that $\Delta v/v \sim 7.8\%$, which – in view of the large experimental errors – is in satisfactory agreement with 3.4 ± 2.0% obtained in our experiments.

According to Chynoweth and Schneider [84] the ratio α/f^2 at t = 17.7°C is approximately three times as large as at t = 15°C. Our results indicate that the dispersion of the velocity of sound increases by the same factor, in agreement with Eq. (III.3). Consequently, the increase in the velocity of hypersound and in the dispersion of this velocity near the critical solution point can be attributed to an increase in the bulk viscosity (the bulk-viscosity relaxation time remains practically constant).

TABLE 5. Refractive Index ($n_{6328Å}$), Velocity of Hypersound (v_h), and Dispersion of the Velocity of Sound ($\Delta v/v$) near the Critical Solution Point t_C = 17.9°C of a 44.6 wt.% Solution of Triethylamine in Water

t, °C	$n_{6328Å}$	v_u, m/sec [†]	v_h, m/sec	$\dfrac{\Delta v}{v}$, % [*]
9	1.381	—	1485$\pm$18	—
11	1.380	1410	1469$\pm$10	4.1$\pm$0.7
13	1.379	1391	1423$\pm$7	2.3$\pm$0.5
15	1.378	1366	1413$\pm$27	3.4$\pm$2.0
16	—	1352	1423$\pm$35	5.1$\pm$2.5
16.5	—	1343	1463$\pm$43	8.6$\pm$3.1
17	1.377	1334	1468$\pm$44	9.6$\pm$3.1
17.3	—	1327	1486$\pm$44	11.3$\pm$3.1
17.6	—	1321	1566$\pm$72 [‡]	17.0$\pm$5.0
17.7	—	1319	1513$\pm$27	13.7$\pm$1.9
—	—	1386 [§]	—	8.8$\pm$1.9
17.8	1.377	1317	1538$\pm$92	15.5$\pm$6.4

* $v = \dfrac{1}{2}(v_h + v_u)$; $\Delta v = v_h - v_u$.

† Velocity of ultrasound calculated from a graph given in [84].

‡ In two (out of many) series of measurements at this temperature the velocity of hypersound was v_h = 1340 ± 40 m/sec.

§ Two values of the velocity of ultrasound (Fig. 17) are given for this temperature in [84].

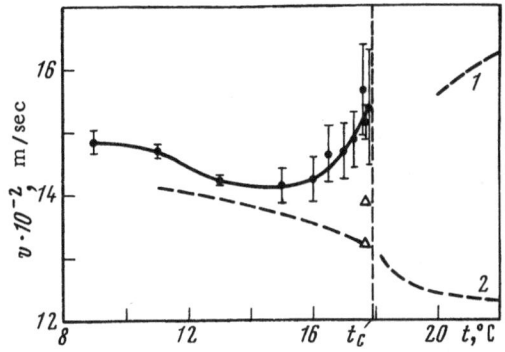

Fig. 17. Velocity of hypersound near the critical solution point of a 44.6 wt.% solution of triethylamine in water (continuous curve). The dashed curve represents the velocity of ultrasound under the same conditions [84] (1 is the phase consisting of water and 2 is the phase consisting of triethylamine).

Fig. 18. Dispersion of the velocity of sound near the critical solution temperature of a 44.6 wt.% solution of triethylamine in water.

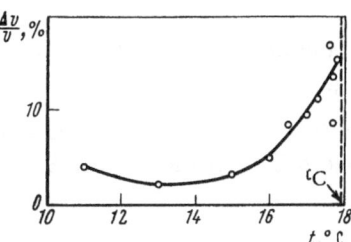

Chen and Polonsky [28] determined the fine structure of the Rayleigh line in the critical solution region of a solution containing 0.42 molar fractions of nitrobenzene in normal hexane. This solution has an upper critical temperature $t_C = 20.20 \pm 0.10°C$. Chen and Polonsky recorded the spectra by the photoelectric method. They found that when the critical point was approached from above (the temperature t was lowered), the shift of the Brillouin components first increased weakly and then — in the range $(t - t_C) \leq 1$ deg — it decreased weakly. Chen and Polonsky did not determine the temperature dependence of the velocity of hypersound because they did not measure the refractive index of the solution. However, if we assume that the refractive index does not vary greatly with the temperature, we can use the shift of the Brillouin components in place of the temperature dependence of the velocity of hypersound. Then, the temperature dependence of this velocity deduced from the data of Chen and Polonsky in the critical region of a solution with an upper critical point is found to be opposite to the temperature dependence of the velocity of hypersound that we obtained for the critical region of a solution with a lower critical point. Chen and Polonsky [28] did not give any data on the velocity and absorption of ultrasound in the solution that they investigated and therefore it is not possible to carry out an analysis of their results of the type described above.

The results reported in [28, 79, 80] should stimulate further studies of the fine structure of the Rayleigh line of solutions near their critical solution point.

CHAPTER V

STIMULATED SCATTERING OF LIGHT IN CARBON DISULFIDE AND IN SOLUTIONS OF CARBON DISULFIDE IN CARBON TETRACHLORIDE. AMPLIFICATION OF THE R_2-LINE RADIATION OF A RUBY LASER BY FOUR-PHOTON INTERACTION

§1. Stimulated Concentration Scattering. Estimate of the Gain Deduced from the Steady-State Theory

In this section we shall consider the steady-state theory of the stimulated concentration scattering. We shall derive an expression for the gain experienced by concentration waves and we shall consider the possibility of experimental observation of the stimulated concentration scattering.

The total free energy per unit mass of a binary solid solution subjected to an external electric field **E** is of the form [38, 43]

$$\varphi = \varphi_0(p, T, c) - \frac{\chi V}{2} E^2, \tag{V.1}$$

where the electric susceptibility χ is

$$\chi = \frac{\varepsilon - 1}{4\pi}. \tag{V.2}$$

The differential $d\varphi$ is given by

$$d\varphi = -s\,dT + V\,dp + \mu_1 dn_1 + \mu_2 dn_2 - \mathbf{P}d\mathbf{E}. \tag{V.3}$$

Here, n_1 and n_2 are the numbers of molecules of both substances per gram of solution; μ_1 and μ_2 are the chemical potentials of these substances; $\mathbf{P} = \chi\mathbf{E}$ is the polarization of the solution. The numbers n_1 and n_2 satisfy the relationship

$$n_1 m_1 + n_2 m_2 = 1, \tag{V.4}$$

where m_1 and m_2 are the masses of the molecules of each substance. If the variable quantity is the concentration $\widetilde{c} = n_1 m_1$, we find that

$$d\varphi = -s\,dT + V\,dp + \mu\,d\widetilde{c} - \mathbf{P}d\mathbf{E}, \tag{V.5}$$

where the chemical potential μ is

$$\mu = \frac{\mu_1}{m_1} - \frac{\mu_2}{m_2}. \tag{V.6}$$

The change in the chemical potential $\delta\mu$ resulting from the application of a field E is

$$\delta\mu = \left(\frac{\partial(\varphi - \varphi_0)}{\partial\widetilde{c}}\right)_{p, T} = -\frac{E^2}{2}\left(\frac{\partial(\chi V)}{\partial\widetilde{c}}\right)_{p, T} \tag{V.7}$$

Since the volumes of the molecules are approximately equal, it follows that

$$\frac{\partial\,(\chi V)}{\partial\widetilde{c}} = V\,\frac{\partial\chi}{\partial\widetilde{c}} + \chi\,\frac{\partial V}{\partial\widetilde{c}} \approx V\,\frac{\partial\chi}{\partial\widetilde{c}} = \frac{1}{4\pi}\,V\,\frac{\partial\varepsilon}{\partial\widetilde{c}}\,. \tag{V.8}$$

Hence,

$$\delta\mu = -\,\frac{E^2}{8\pi}\,V\left(\frac{\partial\varepsilon}{\partial\widetilde{c}}\right)_{p,\,T}. \tag{V.9}$$

If the deviations of the chemical potential from its average value are small, the time and space dependences of this potential are given by the diffusion equation

$$\frac{\partial\mu}{\partial t} - D\nabla^2\mu = \delta\dot{\mu}. \tag{V.10}$$

The rest of the treatment is exactly the same as the analysis given by Starunov [87] for the case of the stimulated temperature scattering.

Let us consider simultaneously the right-hand side of the diffusion equation [Eq. (V.10)] and the nonlinear Maxwell equations in which the nonlinear correction to the polarization of the medium is

$$\mathbf{P}^{nl} = \frac{1}{4\pi}\,\frac{\partial\varepsilon}{\partial\mu}\,\delta\mu\mathbf{E}. \tag{V.11}$$

We shall assume that three plane linearly polarized light waves are traveling in a medium and that their total field is E given by

$$E = \frac{1}{2}\sum_{l=0}^{2} E_l\exp\left(i\omega_l t - i\mathbf{k}_l\mathbf{r}\right) + \text{complex conjugate} \tag{V.12}$$

where the subscripts 0, 1, and 2 represent the laser, Stokes, and anti-Stokes radiation, respectively. We shall also assume that

$$|E_0|\gg|E_1|,\,|E_2|,\quad \omega_0 = \omega_1 + \Omega = \omega_2 - \Omega\ \text{and}\ \Omega\ll\omega_0.$$

It follows from the diffusion equation that the change in the chemical potential of the solution $\delta\mu$ is given by the following expressions:

$$\delta\mu = -\,\frac{\left(\dfrac{\partial\varepsilon}{\partial\widetilde{c}}\right)_{p,\,T}}{16\pi\rho}\left\{\frac{i\Omega E_0 E_1^*\exp\{i\Omega t - i\,(\mathbf{k}_0 - \mathbf{k}_1)\,\mathbf{r}\}}{i\Omega + D\,(\mathbf{k}_0 - \mathbf{k}_1)^2} + \frac{i\Omega E_0^* E_2\exp\{i\Omega t - i\,(\mathbf{k}_2 - \mathbf{k}_0)\,\mathbf{r}\}}{i\Omega + D\,(\mathbf{k}_2 - \mathbf{k}_0)^2}\right.\left.+\,\text{complex conjugate}\right. \tag{V.13}$$

It follows from Eq. (V.13) and from the nonlinear Maxwell equations that the anti-Stokes component should decay and the Stokes component should grow exponentially in space and that the gain is

$$g_c = -\,2k_\omega + B_c|\mathbf{k}_1|\,\frac{\Omega/\Omega_c}{1 + \Omega^2/\Omega_c^2}\,|E_0|^2, \tag{V.14}$$

where $2k_\omega$ is the extinction coefficient; $\Omega_c = D\,(\mathbf{k}_0 - \mathbf{k}_1)^2$, $B_c = \dfrac{\partial\varepsilon}{\partial\widetilde{c}}\,\dfrac{\partial\varepsilon}{\partial\mu}\Big/16\pi n^2\rho$. It is evident from Eq. (V.14) that the Stokes stimulated concentration scattering is strongest at the frequency corresponding to the half-width of the thermal scattering line resulting from fluctuations of the concentration.

We shall now express B_c in the form which is convenient in estimates. We shall start with $\frac{\partial \varepsilon}{\partial \mu} = \frac{\partial \varepsilon}{\partial \tilde{c}} \frac{\partial \tilde{c}}{\partial \mu}$ and

$$\frac{\partial \mu}{\partial \tilde{c}} = \frac{1}{m_1} \frac{\partial \mu_1}{\partial \tilde{c}} - \frac{1}{m_2} \frac{\partial \mu_2}{\partial \tilde{c}}. \tag{V.15}$$

If μ_2 refers to the solvent and μ_1 to the solute (we shall consider the case of weak solutions), it follows from [42] that

$$\left. \begin{array}{l} \mu_2 = \mu_0 - Tc, \\ \mu_1 = T \ln c + \psi, \end{array} \right\} \tag{V.16}$$

where μ_0 is the chemical potential of the pure solvent; T is the temperature in ergs; $c = n_1/n_2$; $n_2 \gg n_1$; ψ is a function of pressure and temperature. If $m_1 \approx m_2$, then $c \approx \tilde{c}$ in respect of the absolute values (c is the dimensionless concentration and $\tilde{c}$ is expressed in grams). Then, substituting the expressions in Eq. (V.16) into Eq. (V.15), we obtain

$$\frac{\partial \mu}{\partial c} \approx \frac{T}{cm_1}. \tag{V.17}$$

Hence,

$$B_c = \frac{\left(\frac{\partial \varepsilon}{\partial c}\right)^2 \frac{cm_1}{T}}{16\pi n^2 \rho}. \tag{V.18}$$

Let us consider the solution of acetone in carbon disulfide discussed in Chap. II. The parameters of this solution are $c \approx 0.13$; $m_2 = M_{C_3H_6O}/N_A \sim 10^{-22}$ g; $\varepsilon = n^2 \approx 1.6^2 = 2.56$; $\partial \varepsilon/\partial c = 2n(\partial n/\partial c) \approx 3.2 \times 5 \times 10^{-1} = 1.6$; $(\partial \varepsilon/\partial c)^2 \sim 2.56$; $\rho = 1.3$ g/cm^3; $T = 3 \times 10^2 \times 1.4 \times 10^{-16} \approx 4 \times 10^{-14}$ erg. The wave vector $|k_0|$ of the exciting radiation ($\lambda \approx 0.7 \times 10^{-4}$ cm) is

$$|k_0| = n \frac{2\pi}{\lambda} \sim 1.5 \cdot 10^5 \text{ cm}^{-1}.$$

We shall estimate the intensity of the electric field of the incident wave by considering a laser beam of ~ 100 MW power, ~ 1 cm diameter, and $\sim 10^{-2}$ rad divergence. When this beam is focused into a solution by a lens of $f = 4$ cm focal length, the area of the focal spot is $\sim 10^{-4}$ cm^2. Thus, the optical energy flux W is $\sim 10^{19}$ erg·sec^{-1}·cm^{-2}. Since $W = ncE_0^2/4\pi$ (c is the velocity of light), it follows that $E_0^2 \approx 2 \times 10^9$ cgs esu and $E_0 \approx 10^7$ V/cm.

We can now estimate the maximum gain of the stimulated concentration scattering. If we assume that the extinction coefficient is small, we find that $g_{c,\max} \approx 750$ cm^{-1}. If l is the length of the region of interaction between the laser and the Stokes waves, the intensity of the Stokes wave at the output, $I_s(l)$, is

$$I_s(l) = I_s(0) e^{g_c l}, \tag{V.19}$$

where $I_s(0)$ is the intensity of the background radiation at the Stokes frequency. If $l \approx 0.1$ cm, the intensity of the Stokes concentration scattering increases by a factor of e^{75}.

A gain by a factor of e^{10}–e^{20} above the background level is usually necessary for the observation of the stimulated scattering [33]. The giant gain by a factor of e^{75} cannot be realized experimentally. This gain by a factor of e^{75} means that one incident photon produces, on the average, 10^{35} new photons. This number is much larger than the initial number of photons in the laser beam, which is about 10^{19} for an energy of 1 J. The intensity of the radiation in the stimulated scattering spectrum cannot exceed the initial intensity of the laser beam and when the former approaches the latter, saturation phenomena occur which are ignored in the theory given above. The high value of the gain, $g_{c,max} \approx 750$ cm^{-1}, simply means that the stimulated concentration scattering can be achieved quite easily under the steady-state conditions discussed above. If the laser field E_0 acts during a time interval Δt, where $\Delta t \ll T_c$ and $T_c = 2\pi/\Omega_c$ is the period of a concentration "wave," the process is transient. The transient nature of the scattering process can be allowed for approximately by multiplying $g_{c,max}$ by $\Delta t/T_c$, i.e., $g_{tr} \approx (\Delta t/T_c)g_{st}$, where the subscripts "tr" and "st" refer to the transient and steady-state conditions [88]. In the case of solutions with large interdiffusion coefficients the period of a concentration "wave" is $T_c \sim 10^{-5}$ sec (Chap. II). Hence, we find that when $\Delta t \sim 10^{-8}$ sec, the relevant ratio is $\Delta t/T_c \sim 10^{-3}$ and $g_{c,max} \approx 0.75$ cm^{-1}.

A gain of e^{10} can be attained by increasing the duration of a laser pulse and at the same time retaining its high average power, i.e., by increasing the energy of the laser radiation. Theoretically, a volume of 1 cm^3 in a crystal can emit radiation of ~ 1 J energy [89]. Light pulses of 6–100 nsec duration can be generated by varying the operating time of optical switches, dimensions of the laser resonator, and the pumping level [90]. If the pulses are of $\Delta t = 100$ nsec duration, the stimulated concentration scattering can be induced if the laser output power is 1 GW and, consequently, the energy of the laser radiation is 100 J. Theoretically, this energy can be generated in a ruby crystal of 1 cm^2 area and 1 m long. However, it is not clear what would happen to a solution that would receive a pulse of this energy. It might happen that the solution would evaporate in the focal region before the stimulated concentration scattering could appear.

The steady-state theory of the stimulated concentration scattering was discussed also by Bespalov and Freidman [91]. Their work has not yet been published and, therefore, it is not possible to compare their results with those given here.

§2. Four-Photon Interaction in Stimulated Scattering Processes

The interaction between the Stokes and anti-Stokes waves is ignored in the theory outlined above. However, in a nonlinear medium the laser, Stokes, and anti-Stokes waves of frequencies ω_0, ω_1, and ω_2, respectively, can interact and this interaction can give rise to a nonlinear polarization of frequency $\omega_0 = \omega_1 + \omega_2 - \omega_0$:

$$P_i^{nl}(\omega_0, \mathbf{k}_p) = \chi_{ijkl}(-\omega_0, \omega_1, \omega_2, -\omega_0) E_j(\omega_1 \mathbf{k}_1) E_k(\omega_2, \mathbf{k}_2) E_l(\omega_0, \mathbf{k}_0), \qquad (V.20)$$

where

$$\mathbf{k}_p = \mathbf{k}_1 + \mathbf{k}_2 - \mathbf{k}_0. \qquad (V.21)$$

The nonlinear polarization $P^{nl}(\omega_0, \mathbf{k}_p)$ is associated with a fourth wave of frequency ω_0 whose wave vector is $\mathbf{k}_p$. This wave is coupled, through the appropriate components of the nonlinear susceptibility, to each of the first three waves. For example, the wave $E(\omega_0, \mathbf{k}_p)$ is coupled to the waves $E(\omega_0, \mathbf{k}_0)$ and $E(\omega_1, \mathbf{k}_1)$ and this coupling produces polarization at the anti-Stokes frequency ω_2, etc. [33]. Consequently, the weak Stokes and anti-Stokes waves may be amplified. The interaction between the waves is strongest if all the waves travel at small angles relative to one another and if $\mathbf{k}_p = \mathbf{k}_0$, i.e., if

$$2\mathbf{k}_L = \mathbf{k}_1 + \mathbf{k}_2. \qquad (V.22)$$

Equation (V.22) shows that two laser photons create Stokes and anti-Stokes photons. The Stokes and anti-Stokes components of the radiation are amplified to the same degree.

The nonlinear coupling between light waves is due to the same nonlinear effects which are responsible for the usual stimulated scattering. In liquids with anisotropic molecules the largest correction to the nonlinear polarization is due to the quadratic Kerr effect which gives rise to stimulated scattering in the wing of the Rayleigh line. The theory of the four-photon interaction in the case of stimulated scattering in the wing of the Rayleigh line has been developed by Starunov [3, 92] and by Chiao, Kelley, and Garmire [34]. Starunov [87] developed also a theory of the four-photon interaction for the case of stimulated temperature scattering. The same approach can be followed in discussing the four-photon interaction in the stimulated concentration scattering.

In the four-photon interaction in the case of stimulated scattering in the wing of the Rayleigh line the nonlinear correction to the refractive index is positive. It is then found that the increase in the refractive index is greater for weak waves. This can be demonstrated most conveniently by following the treatment of Chiao, Kelley, and Garmire [34].

Let us consider the interaction between the two plane waves in a medium with a refractive index which depends on the intensity of light. The total electric field of these two waves is $E = E_0 + E_1$, where

$$E_i = \frac{1}{2}\{\mathscr{E}_i \exp[i(\mathbf{k}_i\mathbf{r} - \omega_i t)] + \text{complex conjugate } (i = 0, 1).$$

Let us assume that E_0 is a strong laser wave and E_1 is a weak scattered-light wave. For simplicity, we shall consider the degenerate case when $\omega_0 = \omega_1$. The intensity-dependent component of the permittivity is

$$\Delta\varepsilon = \varepsilon_2 E^2 = \frac{1}{2}\varepsilon_2[|\mathscr{E}_0|^2 + (\mathscr{E}_0\mathscr{E}_1^* e^{i\mathbf{qr}} + \text{complex conjugate})],$$

where $\varepsilon_2 > 0$, $\mathbf{q} = \mathbf{k}_0 - \mathbf{k}_1$, and the terms containing $\mathscr{E}_1^2$ as well as those oscillating at a frequency $2\omega_0$ are ignored.

The nonlinear correction to ε is responsible for a wave with nonlinear polarization which is given by (in the first-order approximation with respect to the weak wave)

$$\Delta P = \Delta\varepsilon E = (\varepsilon_2/8\pi)\{|\mathscr{E}_0|^2(E_0 + 2E_1) + \frac{1}{2}[\mathscr{E}_0^2\mathscr{E}_1^* \exp\{i(\mathbf{k}_0 + \mathbf{q})\mathbf{r} - i\omega_0 t\} + \text{complex conjugate}]\}$$

We can see that the coefficient associated with the weak wave E_1 is half the coefficient associated with the strong wave E_0. It means that the refractive index for the weak wave increases by Δn compared with the corresponding refractive index for the strong wave. We then find that

$$\Delta n = \Delta n_w - \Delta n_s = \varepsilon_2|\mathscr{E}_0|^2/4n_0,$$

where n_0 is the refractive index of the medium in a weak optical field.

Fig. 19. Phase-matching condition for the four-photon
interaction in the stimulated scattering corresponding
to the wing of the Rayleigh line.

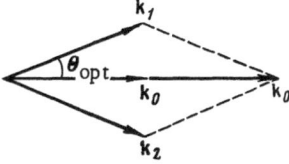

We can say that the strong laser radiation transforms the medium so that it acquires a
large refractive index in respect of the weak wave. The Stokes and anti-Stokes waves are
slowed down. The wave vectors of these waves become longer. Consequently, in order to satisfy the law of conservation of momentum given by Eq. (V.22), these wave vectors should be
oriented at an angle θ_{opt} with respect to the direction of the laser beam (Fig. 19). According
to the theory given in [34, 92], the angle θ_{opt} is

$$\theta_{opt} = (A \, | \, E_0 \, |^2)^{1/2}. \tag{V.23}$$

The coefficient A will be explained later. It follows from the above equation that the higher the
power of the laser radiation, the greater is the angle of the four-photon interaction θ_{opt}. Estimates show that in fields $E_0 \sim 10^7$ V/cm the angle θ_{opt} for liquids with anisotropic molecules
is of the order of several degrees. In the four-photon interaction in the wing of a Rayleigh line
the amplitudes of the Stokes and anti-Stokes waves grow exponentially in space and this growth
is represented by the gain [34, 92]

$$g = - 2k_{\omega} + \frac{A \, | \, \mathbf{k}_0 \, | \, | \, E_0 \, |^2}{(1 + \Omega^2 \tau^2)^{1/2}}, \tag{V.24}$$

where τ is the molecular reorientation time (the anisotropy relaxation time) and

$$A = \frac{8\pi}{45} \frac{N \, (\alpha_1 - \alpha_2)}{\varepsilon_0 kT}. \tag{V.25}$$

Here, N is the number of anisotropic molecules per unit volume; α_1, $\alpha_2 = \alpha_3$ are the principal
polarizabilities of the molecules (it is assumed that the molecules have a symmetry axis along
which the polarizability is strongest); k is the Boltzmann constant.

It is evident from Eq. (V.24) that the highest value of the gain corresponds to $\Omega = 0$, i.e.,
to the case when all four waves have the same frequency ω_0. This case, known as the degenerate four-photon interaction, was first observed experimentally by Carman, Chiao, and Kelley
[93]. However, even if $\Omega > 0$, the gain g can be quite large. We shall estimate gain of the Stokes
and anti-Stokes waves of frequencies $\omega_0 - \Omega$ and $\omega_0 + \Omega$, respectively, where $\Omega \sim 5 \times 10^{12}$ rad/sec
($\Omega/2\pi c \sim 29$ cm^{-1}), which occurs in the four-photon interaction in the wing of the stimulated
Rayleigh line in carbon disulfide. It is known that carbon disulfide is a liquid which consists
of strongly anisotropic molecules. The necessary parameters for this case are: $\rho = 1.26$ g/cm^3;
$M_{CS_2} = 76$; $N = \rho N_A / M_{CS_2} \sim 10^{22}$; $\varepsilon_0 = n^2 \sim 3$; $\alpha_1 \sim 150 \times 10^{-25}$ cm^3; $\alpha_2 \sim 50 \times 10^{-25}$ cm^3 [94];
$(\alpha_1 - \alpha_2) \sim 10^{-23}$ cm^3; kT $\sim 4 \times 10^{14}$ ergs; $| \, k_0 \, | \sim 1.5 \times 10^5$ cm^{-1}; $\tau \sim 2.4 \times 10^{-12}$ sec [1]; $| \, E_0 \, |^2 \sim$
2×10^9 cgs esu ($| \, E_0 \, | \sim 10^7$ V/cm). A calculation in which these parameters are used gives $g \approx$
80 cm^{-1}. The four-photon interaction angle is $\theta_{opt} = 9 \times 10^{-2}$ rad ($\sim 5°$).

The four-photon interaction corresponding to $\Omega > 0$ was first observed by Zaitsev, Kyzylasov, Starunov, and Fabelinskii [95]. They studied the stimulated scattering of ruby laser radiation in nitrobenzene and they observed the anti-Stokes wing of the Rayleigh line extending up to
1.5-2 cm^{-1}.

So far we have considered the Stokes and anti-Stokes radiation which is present in a me-

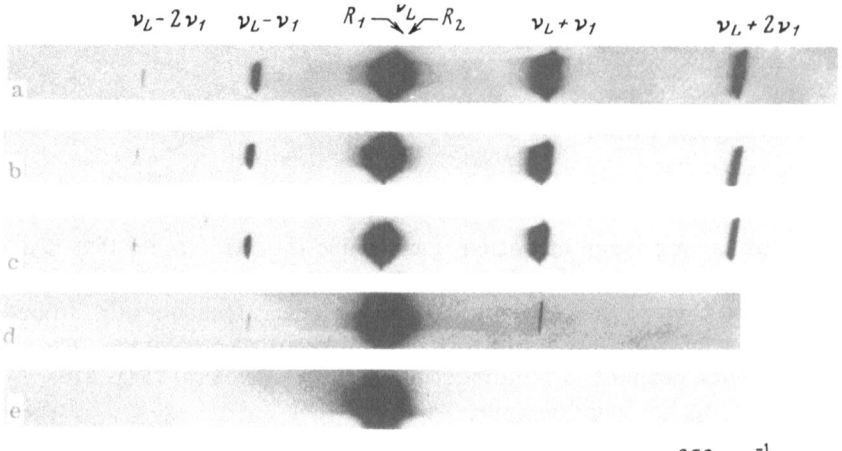

$\nu_1 = 656 \ \mathrm{cm}^{-1}$

Fig. 20. Spectra of the stimulated scattering of light in solutions of carbon disulfide in carbon tetrachloride with the following concentrations of CCl_4 (vol.%): a) 0; b) 25; c) 50; d) 75; e) 90. The spectra were recorded using an ISP-51 spectrograph with a linear dispersion of 14 Å/mm in the region of the R_1 ruby line.

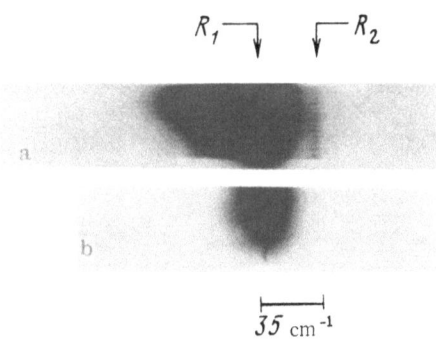

$\overline{35} \ \mathrm{cm}^{-1}$

Fig. 21. Spectrum of the stimulated scattering of light near the R_1 line of ruby: a) carbon disulfide; b) carbon tetrachloride.

dium in the form of noise at the appropriate frequencies. Let us now assume that strong laser beam of frequency ω_0 and a weak beam of frequency $\omega_0 + \Omega$ or $\omega_0 - \Omega$ reaches a nonlinear medium. If some of the weak radiation is focused by a lens so that it meets the laser beam at the four-photon interaction angle and if the phase relationships are correct, the weak radiation can be amplified by the four-photon interaction effect. This effect was observed in the experiments of Aref'ev et al. [96], which will be discussed below.

§3. Stimulated Scattering in Carbon Disulfide and in Solutions of Carbon Disulfide in Carbon Tetrachloride

Carbon disulfide is a liquid which is often employed in demonstrations of stimulated scattering of light. Light is self-focused easily in this liquid. The critical power necessary for such self-focusing is 16 kW [33]. The intensity of the radiation increases tens of times in the self-focusing filaments and various types of stimulated scattering are then possible.

Self-focusing of light is due to the degenerate four-photon interaction [34, 93, 97]. A laser beam of finite diameter can be represented by a superposition of plane waves with different wave vectors **k**. Scattering produces waves with different wave vectors **k**. However, exponential growth is experienced only by those components whose wave vectors are directed at the four-photon interaction angle with respect to k_0. At low laser radiation powers this angle is small and light is not scattered through large angles.

Figures 20 and 21 show the stimulated scattering spectra of carbon disulfide at 19°C which were obtained using the R_1 ruby laser line and spectrographs with small and large dispersion. These spectra were recorded with apparatus shown in Fig. 10. A laser beam of ~ 100 MW/cm^2 power, emitted in the form of pulses of ~ 10 nsec duration and 10^{-2} rad divergence, was focused by an $f = 4$ cm lens into a cuvette containing carbon disulfide. A second lens was placed behind the cuvette and the two lenses together formed an approximately telescopic system. The light collected by the second lens was directed onto the slit of a spectrograph. The spectra were photographed on a 10N-1000 panchromatic film. Figure 20 shows the spectra obtained with an ISP-51 prism spectrograph with a linear dispersion of 29 cm$^{-1} \cdot$mm^{-1} in the region of R_1 line of $\lambda = 6943$Å (use was made of the autocollimation variant of the spectrograph with a camera of $f = 1300$ mm). The spectra shown in Fig. 21 were obtained with the aid of a DFS-8 diffraction spectrograph with a linear dispersion of 3.5 cm$^{-1} \cdot$mm^{-1} (a diffraction grating of 1800 lines/mm was used).

It is evident from Fig. 21 that the investigated part of the scattering spectrum of carbon disulfide, extending from 12,900 to 15,900 cm^{-1}, included not only the laser line at 14,403 cm^{-1} but also the stimulated wing of the Rayleigh line extending to ~ 150 cm^{-1} and four lines of the stimulated Raman scattering (two Stokes and two anti-Stokes) associated with the fundamental vibration of the carbon disulfide molecule $\nu_1 = 656$ cm^{-1}. The second harmonics appear because of the local multiple scattering: the scattered light is sufficiently strong to produce another scattering component. It is interesting to note that all the stimulated Raman scattering lines have a strong stimulated Stokes wing extending up to ~ 100 cm^{-1}.

It seemed interesting to determine the influence of dilution of carbon disulfide with carbon tetrachloride on the stimulated scattering spectrum. Molecules of carbon tetrachloride are spherically symmetrical. Consequently, the four-photon interaction, which is due to the same mechanism as the stimulated wing of the Rayleigh line, is not observed in carbon tetrachloride. Solutions with volume concentrations of 25, 50, 75, and 90% of CCl_4 in CS_2 (corresponding to 0.10, 0.38, 0.75, and 0.80 molar fractions, respectively) were prepared. The concentration of 0.10 molar fractions of CCl_4 indicated that for every ten molecules of CCl_4 there were ninety molecules of CS_2, etc.

It is evident from Fig. 20 that dilution of CS_2 with 50% CCl_4 did not alter greatly the stimulated scattering spectrum. When the concentration of CCl_4 was increased to 75%, only the Stokes and anti-Stokes component of the stimulated Raman scattering line ($\nu_1 = 656$ cm^{-1}) remained in the spectrum. The spectrum of the solution with 90% CCl_4 consisted simply of the laser line.

These results showed that studies of the stimulated scattering in solutions can be used to determine the predominant nonlinear process in the pure liquid. In the case of carbon disulfide our results showed that the stimulated Raman scattering is the first to appear. This makes it possible to interpret with assurance the stimulated Stokes wings of the Raman lines. The results obtained suggest that these wings are due to excitation of the stimulated wing of the Rayleigh line by the stimulated Raman line and not the other way round.

§4. Amplification of the R_2 Ruby Laser Line by the Four-Photon Interaction in the Stimulated Wing of the Rayleigh Line

The stimulated scattering spectra of carbon disulfide included, apart from the Raman line and the wing of the Rayleigh line, an additional line which was displaced in the anti-Stokes

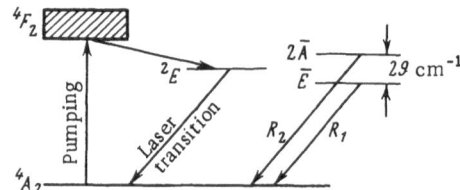

Fig. 22. Energy level scheme of ruby [98].

direction by 29 cm^{-1} from the R_1 ruby laser line. The frequency of this line was 14,432 cm^{-1} and it coincided with the R_2 ruby laser line. This additional line was not observed in the laser radiation spectrum or in the spectra of the stimulated scattering in carbon tetrachloride and acetone. This line disappeared in the solutions of CS_2 in CCl_4 (Fig. 20) if the stimulated scattering in the wing of the Rayleigh line was not observed.

An investigation carried out with the high-dispersion spectrograph (DFS-8) made it possible to study this line in greater detail. The spectra obtained with this spectrograph (Fig. 21) included also the anti-Stokes wing of the Rayleigh line, which was indistinguishable from the background in the spectra recorded with the ISP-51 spectrograph (Fig. 20). Moreover, the stimulated scattering spectra of carbon disulfide (Fig. 21) included also narrow bands on the Stokes and anti-Stokes sides of the R_1 laser line, which corresponded to the Brillouin components not resolved by the spectrograph. The width of the band on the Stokes side was ~ 1 mm. This meant that the spectrum contained approximately 20 Stokes component of the Brillouin scattering.

These results showed that the appearance of the anti-Stokes line at 14,432 cm^{-1} in the stimulated scattering spectrum of carbon disulfide is due to the amplification of the weak R_2 laser line as a result of the four-photon interaction in the stimulated wing of the Rayleigh line.

The energy levels of ruby ($Al_2O_3 : Cr^{3+}$) are shown schematically in Fig. 22. At low concentrations of chromium the Cr^{3+} ion, which experiences a field of cubic symmetry, should have a single resonance line corresponding to the $^2E \rightarrow {}^4A_2$ transition responsible for the laser emission. In fact, the chromium ions occupy somewhat distorted octahedra consisting of oxygen ions and the ^2E level splits into the $2\bar{A}$ and $\bar{E}$ levels, which are separated by 29 cm^{-1} [98-100]. The $2\bar{A} \rightarrow {}^4A_2$ transition gives rise to the R_2 line whereas the $\bar{E} \rightarrow {}^4A_2$ transition produces the R_1 line. Almost all the available information on the applications of ruby radiation is concerned with the R_1 line. McClung, Schwarz, and Meyers [101] generated the R_2 line by inserting, in the laser resonator, multilayer dielectric mirrors with a high transmission coefficient for the R_1 line and a high reflection coefficient for the R_2 line.

In our investigation no special measures were taken to ensure the emission of the R_2 line. Our laser emitted either a fairly strong spontaneous radiation or weak stimulated radiation at the R_2-line frequency. The enhancement of the R_2 ruby line was not observed in some of the ruby crystals. In such cases the R_2 line was absent from the emission of the Q-switched laser or it was so weak that even amplification by a factor g ~ 10^2 failed to reveal it in the scattering spectrum. On the other hand, in some of the earlier experiments [31] the R_2 line was observed even in the unamplified laser radiation. The mechanism of the emission of the R_2 line is of intrinsic interest but it is outside the scope of the present paper. The R_2 line is not observed in the scattering spectra of carbon tetrachloride and acetone because in the case of CCl_4, $\alpha_1 = \alpha_2 = \alpha_3$ and g = 0 and in the case of acetone $\alpha_1 \approx \alpha_2 \sim \alpha_3$ and the gain g is very small.

The possibility of using the four-photon interaction mechanism in the amplification of weak radiation was confirmed by Kyzylasov and Starunov [102]. They observed amplification of the Stokes and anti-Stokes component of the stimulated Brillouin scattering in fused quartz when the scattered radiation was passed together with the laser beam through a cuvette containing carbon disulfide. In this case carbon disulfide served as a nonlinear medium in which the four-photon interaction occurred between the laser radiation and the Stokes and anti-Stokes components of the Brillouin-scattered radiation.

The author is deeply grateful to his scientific supervisor I. L. Fabelinskii for suggesting the subject, directing the investigation, and his constant interest. The author is also indebted to V. S. Starunov for his advice and help in the investigation, to V. P. Zaitsev, G. I. Zaitsev, S. V. Krivokhizha, Yu. I. Kyzylasov, and E. V. Tiganov for their help in the experiments, to D. I. Mash, B. D. Kopylovskii, and I. V. Shtranikh for their assistance in the optical heterodyne experiments.

LITERATURE CITED

1. I. L. Fabelinskii, Molecular Scattering of Light, Plenum Press, New York (1968).
2. V. S. Starunov, Dokl. Akad. Nauk SSSR, 153:1055 (1963); Opt. Spektrosk., 18:300 (1965).
3. V. S. Starunov, Tr. Fiz. Inst. Akad. Nauk SSSR, 39:151 (1967).
4. V. S. Starunov, E. V. Tiganov, and I. L. Fabelinskii, ZhETF Pis. Red., 5:317 (1967).
5. I. L. Fabelinskii and V. S. Starunov, Appl. Opt., 6:1793 (1967).
6. E. V. Tiganov, Dissertation [in Russian], Physics Institute Academy of Sciences of the USSR, Moscow (1967) [Tr. Fiz. Inst. Akad. Nauk SSSR, 58:42 (1972)]. This volume, page 39.
7. G. I. A. Stegeman and B. P. Stoicheff, Phys. Rev. Lett., 21:202 (1968).
8. M. A. Leontovich, J. Phys. USSR, 4:499 (1941).
9. S. M. Rytov, Zh. Eksp. Teor. Fiz., 33:514, 671 (1957).
10. G. S. Gorelik, Dokl. Akad. Nauk SSSR, 58:45 (1947).
11. A. T. Forrester, R. A. Gudmundsen, and P. O. Johnson, Phys. Rev., 99:1691 (1955).
12. H. Z. Cummins, N. Knable, and Y. Yeh, Phys. Rev. Lett., 12:150 (1964).
13. S. B. Dubin, J. H. Lunacek, and G. B. Benedek, Proc. Nat. Acad. Sci. USA, 57:1164 (1967).
14. S. S. Alpert, Y. Yeh, and E. Lipworth, Phys. Rev. Lett., 14:486 (1965).
15. J. A. White, J. S. Osmundson, and B. H. Ahn, Phys. Rev. Lett., 16:639 (1966).
16. B. Chu, Phys. Rev. Lett., 18:200 (1967).
17. N. C. Ford, Jr., and G. B. Benedek, Phys. Rev. Lett., 15:649 (1965).
18. S. S. Alpert, D. Balzarini, R. Novick, L. Seigel, and Y. Yeh, Physics of Quantum Electronics (Proc. Intern. Conf. San Juan, Puerto Rico, 1965), publ. by McGraw-Hill, New York (1966), p. 253.
19. J. B. Lastovka and G. B. Benedek, Phys. Rev. Lett., 17:1039 (1966).
20. K. Sunanda Bai, Proc. Indian Acad. Sci., A18:210 (1943).
21. Kh. E. Sterin, Dokl. Akad. Nauk SSSR, 62:219 (1948).
22. L. V. Lanshina and M. I. Shakhparonov, in: Critical Phenomena and Fluctuations in Solutions [in Russian], Izd. AN SSSR, Moscow (1960), p. 77.
23. L. V. Lanshina, Yu. G. Shoroshev, and M. I. Shakhparonov, Dokl. Akad. Nauk SSSR, 173:70 (1967).
24. I. M. Aref'ev, V. S. Starunov, and I. L. Fabelinskii, ZhETF Pis. Red., 6:677 (1967).
25. Yu. G. Shoroshev, M. I. Shakhparonov, and L. V. Lanchina, Vestn. Mosk. Univ., Khim., 22(5):147 (1967).
26. C. J. Burton, J. Acoust. Soc. Amer., 20:186 (1948).
27. L. I. Mandel'shtam and M. A. Leontovich, ZhETF, 7:438 (1937).
28. S. H. Chen and N. Polonsky, Phys. Rev. Lett., 20:909 (1968).

29. E. J. Woodbury and W. K. Ng, Proc. IRE, 50:2367 (1962).

30. R. Y. Chiao, C. H. Townes, and B. P. Stoichevv, Phys. Rev. Lett., 12:592 (1964).

31. D. I. Mash, V. V. Morozov, V. S. Starunov, and I. L. Fabelinskii, ZhETF Pis. Red., 2:41 (1965).

32. G. I. Zaitsev, Yu. I. Kyzylasov, V. S. Starunov, and I. L. Fabelinskii, ZhETF Pis. Red., 6:802 (1967).

33. N. Bloembergen, Amer. J. Phys., 35:989 (1967).

34. R. Y. Chiao, P. L. Kelley, and E. Garmire, Phys. Rev. Lett., 17:1158 (1966).

35. S. G. Rautian, Usp. Fiz. Nauk, 66:475 (1958).

36. A. T. Forrester, in: Lasers [in Russian], IL, Moscow (1963), p. 289.

37. A. T. Forrester, J. Opt. Soc. Amer., 51:253 (1961).

38. L. D. Landau and E. M. Lifshitz, Fluid Mechanics, Pergamon Press, London (1959).

39. M. A. Leontovich, Zh. Eksp. Teor. Fiz., 49:1624 (1965).

40. E. E. Khazanova and L. A. Rott, Inzh. Fiz. Zh., 6(11):123 (1963).

41. S. S. Alpert, Proc. Conf. on Phenomena in the Neighborhood of Critical Points, Washington, 1965, publ. by National Bureau of Standards, Washington (1966), p. 157.

42. I. R. Krichevskii, N. E. Khazanova, and L. R. Lipshits, Dokl. Akad. Nauk SSSR, 141:397 (1961).

43. L. D. Landau and E. M. Lifshitz, Electrodynamics of Continuous Media, Pergamon Press, Oxford (1960).

44. L. D. Landau and E. M. Lifshitz, Statistical Physics, 2nd ed., Pergamon Press, Oxford (1969).

45. N. C. Ford, Jr., and G. B. Benedek, Proc. Conf. on Phenomena in the Neighborhood of Critical Points, Washington, 1965, publ. by National Bureau of Standards, Washington (1966), p. 150.

46. D. A. Jennings and H. Takuma, Appl. Phys. Lett., 5:241 (1964).

47. R. G. Brewer, Appl. Phys. Lett., 9:51 (1966).

48. H. Takuma and D. A. Jennings, Appl. Phys. Lett., 4:185 (1964).

49. I. M. Aref'ev, B. D. Kopylovskii, D. I. Mash, and I. L. Fabelinskii, ZhETF Pis. Red., 5:438 (1967).

50. A. Javan, E. A. Ballik, and W. L. Bond, J. Opt. Soc. Amer., 52:96 (1962).

51. D. R. Herriott, J. Opt. Soc. Amer., 52:31 (1962).

52. V. V. Sharonov, Light and Color [in Russian], Fizmatgiz, Moscow (1961).

53. Technical Encyclopedia (Handbook of Physical, Chemical, and Technological Data) [in Russian], Vol. 4, OGIZ RSFSR, Moscow (1930), pp. 91-119.

54. Technical Encyclopedia (Handbook of Physical, Chemical, and Technological Data) [in Russian], Vol. 7, OGIZ RSFSR, Moscow (1931), p. 262.

55. J. Ducloux, Diffusion dans les Liquides, Herrmann and Cie., Paris (1938).

56. W. Jost, Diffusion in Solids, Liquids, and Gases, Academic Press, New York (1952).

57. O. Nomoto, J. Phys. Soc. Jap., 11:827 (1956).

58. O. Nomoto, J. Phys. Soc. Jap., 12:300 (1957).

59. V. P. Romanov and V. A. Solov'ev, Akust. Zh., 11:84 (1965).

60. J. M. Davenport, J. F. Dill, V. A. Solov'ev, and K. Fritsch, Akust. Zh., 14:288 (1968).

61. D. I. Mash, V. S. Starunov, E. V. Tiganov, and I. L. Fabelinskii, Zh. Eksp. Teor. Fiz., 49:1764 (1965).

62. S. Tolansky, High Resolution Spectroscopy, Methuen, London (1947).

63. G. I. Zaitsev, Yu. I. Kyzylasov, V. S. Starunov, and I. L. Fabelinskii, ZhETF Pis. Red., 6:505 (1967).

64. G. I. Zaitsev, Dissertation [in Russian], Physics Institute, Academy of Sciences of the USSR, Moscow (1967) [Tr. Fiz. Inst. Akad. Nauk SSSR, 58:3 (1972)]. This volume, page 1.

65. V. V. Morozov, Prib. Tekh. Eksp., No. 2, p. 179 (1968).

66. D. I. Mash, V. V. Morozov, V. S. Starunov, and I. L. Fabelinskii, Zh. Eksp. Teor. Fiz., 55:2053 (1968).

67. T. S. Velichkina, and I. L. Fabelinskii, Dokl. Akad. Nauk SSSR, 75:177 (1950).

68. S. V. Krivokhizha and I. L. Fabelinskii, Zh. Eksp. Teor. Fiz., 50:3 (1966).

69. E. V. Tiganov, ZhETF Pis. Red., 4:385 (1966).

70. M. A. Leontovich, Izv. Akad. Nauk SSSR, Ser. Fiz., No. 5, p. 633 (1936).

71. I. G. Mikhailov, V. A. Solov'ev, and Yu. P. Syrnikov, Fundamentals of Molecular Acoustics [in Russian], Nauka, Moscow (1964).

72. F. Barocchi, M. Mancini, and R. Vallauri, Nuovo Cimento, 49:233 (1967).

73. V. I. Bespalov and A. M. Kubarev, ZhETF Pis. Red., 6:500 (1967).

74. D. I. Mash, V. V. Morozov, V. S. Starunov, E. V. Tiganov, and I. L. Fabelinskii, ZhETF Pis. Red., 2:246 (1965).

75. A. A. Chaban, ZhETF Pis. Red., 3:73 (1966).

76. V. V. Morozov, Dissertation [in Russian], Physics Institute, Academy of Sciences of the USSR, Moscow (1968) [Tr. Fiz. Inst. Akad. Nauk SSSR, 58:80 (1972)]. This volume, page 75.

77. N. R. Goldblatt and T. A. Litovitz, J. Acoust. Soc. Amer., 41:1301 (1967).

78. V. P. Skripov, in: Critical Phenomena and Fluctuations in Solutions [in Russian], Izd. Akad. Nauk SSSR, Moscow (1960), p. 117.

79. I. M. Aref'ev, ZhETF, Pis. Red., 7:361 (1968).

80. I. M. Aref'ev, in: Current Problems in Physical Chemistry [in Russian], Vol. 5, Moscow State University (1970), p. 204.

81. R. W. Gammon, H. L. Swinney, and H. Z. Cummins, Phys. Rev. Lett., 19:1467 (1967).

82. N. C. Ford, Jr., K. H. Langley, and V. G. Puglielli, Phys. Rev. Lett., 21:9 (1968).

83. B. S. Guberman and V. V. Morozov, Opt. Spektrosk., 22:673 (1967).

84. A. G. Chynoweth and W. G. Schneider, J. Chem. Phys., 19:1566 (1951).

85. D. Sette, Nuovo Cimento, 1:800 (1955).

86. V. K. Semenchenko and E. L. Zorina, Zh. Fiz. Khim., 26:520 (1952).

87. V. S. Starunov, Phys. Lett., 26A:428 (1968).

88. N. M. Kroll, J. Appl. Phys., 36:34 (1965).

89. A. A. Vuylsteke, in: Lasers [in Russian], Mir, Moscow (1966), p. 89.

90. P. G. Kryukov, Dissertation [in Russian], Physics Institute, Academy of Sciences of the USSR, Moscow (1966).

91. V. I. Bespalov and G. I. Freidman, Abstracts of Papers Presented at Third Symposium on Nonlinear Optics, Erevan, 1967 [in Russian].

92. V. S. Starunov, Dokl. Akad. Nauk SSSR, 179:65 (1968).

93. R. L. Carman, R. Y. Chiao, and P. L. Kelley, Phys. Rev. Lett., 17:1281 (1966).

94. M. V. Vol'kenshtein, Molecular Optics [in Russian], Gostekhteoretizdat, Moscow-Leningrad (1951).

95. G. I. Zaitsev, Yu. I. Kyzylasov, V. S. Starunov, and I. L. Fabelinskii, ZhETF Pis. Red., 6:695 (1967).

96. I. M. Aref'ev, I. L. Fabelinskii, Yu. I. Kyzylasov, V. S. Starunov, and G. I. Zaitsev, Phys. Lett., 26A:82 (1967).

97. V. I. Bespalov and V. I. Talanov, ZhETF Pis. Red., 3:471 (1966).

98. A. L. Schawlow, S. Vogel, and L. H. Dulberger, in: Lasers [in Russian], IL, Moscow (1962).

99. A. L. Schawlow, in: Lasers [in Russian], IL, Moscow (1963), p. 51.

100. P. Görlich, H. Karras, G. Kötitz, and R. Lehmann, "Spectroscopic properties of activated laser crystals," Phys. Status Solidi, 5:437 (1964); 6:277 (1964); 8:385 (1965).

101. F. J. McClung, S. E. Schwarz, and F. J. Meyers, J. Appl. Phys., 33:3139 (1962).

102. Yu. I. Kyzylasov and V. S. Starunov, ZhETF Pis. Red., 7:160 (1968).